Predicting from Data

An Alternative Unit
for Representing and
Analyzing Two-Variable Data

GLENCOE
Mathematics Replacement Units

GLENCOE
McGraw-Hill

New York, New York
Columbus, Ohio
Mission Hills, California
Peoria, Illinois

Development of the Unit

Predicting from Data is one of two units developed under the auspices of the **Making Mathematics Accessible to All Project.** The project is a statewide, cooperative effort among Michigan higher education institutions and intermediate and local districts to assist high schools in reshaping their mathematics programs to provide a common core of broadly useful mathematics for *all* students. The project is supported by a grant to Western Michigan University from the Dwight D. Eisenhower Mathematics and Science Education Program. In addition to conducting staff development programs for school-based teams consisting of a pair of teachers, a guidance counselor, and an administrator, and conducting follow-up workshops for teachers, the project has also developed a unit titled *Exploring Data,* which is also published by Glencoe.

Send all inquiries to:
Glencoe/McGraw-Hill
936 Eastwind Drive
Westerville, Ohio 43081

ISBN: 0-02-824208-4(Student Edition)
ISBN: 0-02-824209-2 (Teacher's Annotated Edition)

1 2 3 4 5 6 7 8 9 10 VH/LH-P 03 02 01 00 99 98 97 96 95 94

Table of Contents

Teacher's Handbook

Using the Unit

This Teacher's Annotated Edition includes the Student Edition together with Teacher Notes and suggested solutions. The unit consists of six investigations, which are each made up of activities. Each activity is designed around the following instructional model, which is an adaptation and extension of one used in the Middle Grades Mathematics Project (MGMP). The MGMP is a program developed at Michigan State University to develop mathematics units for instruction in grades 5 through 8. The National Science Foundation funded the program. This model, like the unit, itself, requires new roles for both students and teachers.

Instructional Model

Motivate and Focus	During this first step, the teacher initiates whole-class discussion of the problem situation or question(s) posed at the beginning of each investigation and related activity. This sets the context for the student work to follow. It provides an opportunity to clarify directions for the group activities. The teacher is the director and moderator.
Explore	The next step involves focused problems/questions related to the *Motivate and Focus* situation where investigation leads to gathering data, looking for patterns, and making conjectures. Students work cooperatively in small groups. While groups are working, the teacher circulates from group to group providing guidance and support. This may entail clarifying or asking questions, giving hints, providing encouragement, and drawing group members into the discussion to help groups work more cooperatively. The materials drive the learning, and the teacher is the facilitator.
Share and Summarize	This utilizes whole-class discussion of results found by different small groups. It leads to a whole-class summary of important ideas or to further exploration of a topic if competing perspectives remain. Varying points of view and differing conclusions that can be justified should be encouraged. The teacher is the moderator.
Assess	A group task is used to reinforce initial understanding of the concept or method. The teacher circulates around the room assessing levels of understanding. The teacher is the intellectual coach. Depending on the goals of the activity, there may be additional *Explore, Share and Summarize, Assess* sequences. An overall class discussion and synthesis by the teacher should occur at the end of each investigation.
Related Applications	This involves a set of related or new contexts to which student-developed ideas and methods can be applied. A subset of these additional applications should be assigned for further group work in class and/or for individual work outside of class. Decisions should be based on student performance and the availability of time and technology. Students should exercise some choice of applications to pursue; at times they should be given opportunity to pose their own problems/questions to investigate. The teacher circulates around the room acting as facilitator and coach.
Extensions	Finally, students work on exercises which permit further or deeper/more formal study of the topic under investigation. Every student should have the opportunity to complete one or more of these extensions during the course of a unit.

Grouping Your Students

This unit has been designed for use in classrooms containing students of diverse backgrounds. In order to ensure heterogeneous grouping, you should make the final decision about group assignments based on mathematical performance levels, gender, and cultural backgrounds at the beginning of each investigation. Students can be re-grouped later in an investigation for related applications and extensions.

If your students have not previously experienced small-group cooperative learning, you will need to present some guidelines for group behavior. These should included:

- Listen carefully and with respect to each other and try, whenever possible, to build on the ideas of each other.
- Make sure everyone contributes to the group task and no one person dominates.
- Ask for clarification or help when something is not understood and help others in the group when asked.
- Achieve a group answer or solution for each task.
- Make sure everyone understands the solution before sharing it with the class or before the group goes on to the next task.

Expect that group work will go more smoothly on some days than others. It usually takes about three weeks for students to begin functioning well in groups. From time to time, remind students of the above guidelines for cooperative group behavior.

More information about cooperative learning can be found in Glencoe's *Cooperative Learning in the Mathematics Classroom.*

Assessing Your Students

Instruction and assessment are closely linked. Both the Student Edition and Teacher's Annotated Edition provide tasks for group as well as individual assessment. The Teacher Suggestions and Strategies also provide suggestions for methods of informal assessment of student performance based on observations. The nature of the Student Edition and the small group format are ideally suited to assessment of mathematical thinking, communication, and disposition toward mathematics through student observations.

For more information on performance assessment, see Glencoe's *Alternative Assessment in the Mathematics Classroom.*

Using Journals and Portfolios

Journals and portfolios are important forms of assessment. Students will find journal prompts and suggestions for what to include in a portfolio in the margins of the activities.

Journals A journal is a written account that a student keeps to record what she or he has learned. Journal entries are conducive to thinking about why something has been done. They can be used to record and summarize key topics studied, the student's feelings toward mathematics, accomplishments or frustrations in solving a particular problem or topic, or any other notes or comments the student wishes to make. Keeping a mathematical journal can be helpful in students' development of a reflective and introspective point of view. It also encourages students to have a more thoughtful attitude toward written work and should be instrumental in helping students learn more mathematics. Journals are also an excellent way for students to practice and improve their writing skills.

Portfolios A portfolio is a representative sample of a student's work that is collected over a period of time. The selection of work samples for a portfolio should be done with an eye toward presenting a balanced portrait of a student's achievements. The pieces of work placed in a portfolio should have more significance than other work a student has done. They are chosen as illustrations of a student's best work at a particular point in time. Thus, the range of items selected shows a student's intellectual growth in mathematics over time.

Students may select the products of activities for inclusion. Bear in mind that the actual selection of the items by the students will tell you what pieces of work the students think are significant. In addition, students should reflect upon their selections by explaining why each particular work was chosen.

The following examples illustrate topics that would be appropriate for inclusion in a portfolio.

- a solution to a difficult or nonroutine problem that shows originality of thought
- a written report of an individual project or investigation
- examples of problems or conjectures formulated by the student
- mathematical artwork, charts, or graphs
- a student's contribution to a group report or investigation
- a photo or sketch of physical models or manipulative
- statements on mathematical disposition, such as motivation, curiosity, and self confidence
- a first and final draft of a piece of work that shows student growth

Student Diversity

In the *Predicting from Data* activities, students are asked to interpret data to discuss and support approaches to problems. Students will have to read charts, write reports, defend solutions, and draw conclusions. The students in your class may not have the same level of proficiency required to carry out and produce such high level forms of communication. Reasons of this diversity include lack of proficiency in English, poor knowledge of mathematical terminology, limited exposure to the rules and use of language in mathematical contexts, and lack of background and experience with technical forms of communication.

To foster participation and effective communication by *all* students, we must try to obtain an idea of the students' communication skills. This can be accomplished by using the students' responses to each investigation opener. Listen to the students as they talk about and interpret the task. Who is having difficulties understanding the requirements? Which students cannot identify and explain the components and objectives of the activity?

Second, examine the written work. Which students cannot fully explain the solution? Is this due to a lack of mathematical knowledge or the inability to express thoughts in written or oral form?

Lastly, you may wish to use the following general strategies to address communication diversity issues.

- Spend the initial phase of the activities clarifying terms, expressions, task requirements, and symbols. Encourage students to offer their interpretations of the problems or task components.

- Encourage students to describe ideas and processes orally in their cooperative learning groups. Many communication skills are developed through practice.

- Let students use their first language in small groups if they are unable to fully communicate mathematical ideas in English.

Key to Icons in Margin Column

Icon	Description
FYI	**FYIs** are "fast facts," entertaining tidbits, and fascinating math-related trivia.
Graphing Calculator Activity	*Predicting from Data* contains 8 **Graphing Calculator Activities** that allow students to explore mathematical concepts.
Share and Summarize	These headings suggest class discussion about the results found by different groups.
Portfolio Assessment	A **Portfolio Assessment** suggestion asks students to select items from their work that represent their knowledge.
Journal	A **Journal Entry** appears in many of the activities, giving students the opportunity to keep a written log of their thoughts.

Predicting from Data

Unit Overview

Why This Unit is Important

The overall purpose of this unit is to bring students of all abilities in contact with important mathematical ideas and techniques. *Predicting from Data* initiates the work on representation and analysis of paired data. The scatter plot plays a fundamental role throughout the unit.

Investigation Profile

Investigation 1 The first of the six investigations which make up the unit introduces scatter plots as a means of representing paired data geometrically and, therefore, visually. A set of data of high school female and male athletes is used to provide students with information complex enough to demand visual as well as tabular representations. Through this vehicle, the student is led to identify positive, negative, and little or no association between pairs of variables and to intuitively appreciate the measure of association called the correlation coefficient.

Investigation 2 Students learn that predicting from a scatter plot is risky and that, to improve the reliability of the prediction, the data in the scatter plot need to be summarized. The summarizing tool is the line which is visually fit to the data.

Investigation 3 In this investigation, the summarizing line is fit by the process of identifying the median points of three sections of data and drawing the line determined by those points. The result is the median-fit line, which provides a good summary of the data. Since it is constructed by a specific procedure, it brings more uniformity to the predictions based on the summary lines.

Investigation 4 The median-fit line is used to summarize data that lies on a line. Thus the prediction accuracy improves greatly. The idea of a linear equation is also introduced as a representation of the data previously represented by table and by graph.

Investigation 5 The characteristics of linear functions are explored including the notion of the slope and y-intercept of the graph. Pattern analysis is used by students to develop a procedure to find the slope given two points and how to write the equation of a line when given appropriate data.

Investigation 6 In the final investigation, students are asked to use their skill in finding the equation of a line in data analysis situations. They are asked to find an equation of the median-fit line for a set of data and then to make predictions based on that line and its equation. Finally they are led to develop a procedure for solving for x when y is known.

Unit Outcomes

There are three fundamental outcomes of the work in *Predicting from Data*.

- Students will be able to analyze paired data by collecting, organizing, analyzing, and interpreting data using visual, numerical, and algebraic representations. They will be able to make predictions and justify orally or in writing both the prediction and the means used to make it.

- Students will appreciate the usefulness of graphical, tabular, numerical, and algebraic representations in making sense of real-world data.

- Students will recognize and be able to discuss the limitations of modeling real situations with linear models.

Using Manipulatives and Materials

Cooperative learning activities, which often include the use of manipulatives are the basis for all learning in *Predicting from Data*. Below is a list of the manipulatives and materials suggested for this unit. A list specific to each activity also appears at the beginning of the activity.

Manipulative	Activity	Manipulative	Activity
Tape Measure	1-2, 6-1	Ruler	2-1, 2-2, 3-1, 4-1, 4-2
Graph Paper	1-2, 1-3, 2-1, 2-2, 4-1, 4-2, 5-1, 5-2, 5-3, 6-1	Tracing Paper	3-1
Cylindrical Objects	1-2	Calculator	3-1, 4-1, 4-2, 5-1, 6-1
String	1-2, 2-1, 3-1	Colored Pencils	5-2
Software	1-4, 3-1, 4-2, 5-1, 5-2, 6-1, 6-2	Cups and Counters	6-2

● **Using** *Predicting from Data*
with
Merrill Pre-Algebra: A Transition to Algebra
Merrill Algebra 1: Applications and Connections,
Merrill Geometry: Applications and Connections, and
Merrill Algebra 2 With Trigonometry: Applications and Connections

Predicting from Data consists of six investigations that may be used as alternative or supplemental materials for *Merrill Pre-Algebra, Merrill Algebra 1, Merrill Geometry* and *Merrill Algebra 2.* The following correlation shows you which investigation can be used with lessons in each text.

Predicting from Data Investigation Number	*Merrill Pre-Algebra* Lesson Number	*Merrill Algebra 1* Lesson Number	*Merrill Algebra 2* Lesson Number
1	10-6	6-1, 9-2, 14-6	2-1, 2-3, 2-6
2	5-8	9-8, 10-8	2-3
3			2-6, page 91
4	8-6, 8-7	9-4	2-5
5	8-7, 8-8, 8-9	9-3, 10-2, 10-2, 10-3	2-2, 2-4
6	7-2	10-4, 10-5	2-4

Teaching Suggestions and Strategies

Displaying Paired Data

● **Mathematical Overview**

This investigation deals with informal analyses of relations between paired data using tabular representation, pairs of box-and-whisker plots, and scatter plots. From visually reviewing the graphs, students are asked to describe the characteristics of the data and to make predictions for uncharted pairs. The concept of association is introduced to focus the analysis of scatter plots. Technology is used to produce graphs which are then studied for patterns by students.

● **Investigation Outcomes**

Students will
- create a scatter plot for paired data;
- interpret a scatter plot as to whether it displays positive, negative, or no association between a pair of variables;
- interpret a scater plot in terms of the correlation coefficient of paired data;
- use a scatter plot to estimate a value for one variable given a value for the other.

● **Materials Needed**

 string

 graph paper

 tape measure

 cylindrical objects

 software

Activity 1-1   Looking for Relationships

1 Motivate and Focus

Ask students to recall the measures of center and how they are used.

Ask, "Why would Athletic Director Molar want to keep these data?"

Ask, "What methods do we have for displaying data?"

2 Share and Summarize

Bring students together after Exercise 4 to discuss their reasonings and to develop a class consensus. Continue with Exercises 5-9 and develop a class consensus on the questions raised.

3 Extend the Activity

Interview a coach or athletic director at your school to find out what kind of data he/she keeps and how he/she uses that information. What types of data are kept on students at your school and how are they used?

4 Assess the Activity

Refer students to the tables on pages 5 and 6 and ask them to:

a. Prepare side-by-side box-and-whisker plots that compare the GPA's of male and female students in each grade.

b. Determine the mean GPA in each grade for females and for males.

c. Explain the GPA changes as male and female students progress from grade to grade.

d. Determine the median GPA for each grade for each sex.

e. Compare the median with the mean for tenth-grade males and for tenth-grade females and explain any differences.

Activity 1-2   Scatter Plots

1 Motivate and Focus

Ask students, "How do people gather and display data?"

Ask, "Who has ever plotted points on a coordinate system?" Let a student explain how they plot points.

Discuss accuracy of measurements and recording data. If students have completed the unit, *Exploring Data*, they already have the data for Exercise 1a-b.

2 Share and Summarize

Bring students together after Exercise 1 to discuss their data collection processes. Discuss Exercise 5 in detail and talk about positive association. Ask for student summary after Exercise 6. Continue with Exercises 7-14 and discuss the class views regarding "negative association" and "no association." Finish Exercises 15-17 and discuss the class findings.

3 Extend the Activity

Get a list of the top ten songs for your area either from the local radio station or the newspaper. Have a student rank the songs from 1-10 (1 being his/her first choice). Summarize the data on a scatter plot with the media ranking on the horizontal axis and the student ranking on the vertical axis. Ask students, "Are the two rankings associated? Does the student have similar or dissimilar taste in comparison with the station's ranking? Explain."

4 Assess the Activity

On three separate graphs, have students plot the following.

a. 15 data points that show a strong positive association.

b. 15 data points that show a weak positive association.

c. 15 data points that show a strong negative association.

Key ideas can also be assessed with Exercise 18.

Activity 1-3   Using Scatter Plots to Analyze Data

1 Motivate and Focus

Ask students the following questions.

a. What type of technology is available for storing and displaying data?

b. What advantages/disadvantages does technology have for storing and displaying data?

c. What types of patterns can you observe in scatter plots?

2 Share and Summarize

Bring students together after Exercise 5 and discuss the methods for illustrating double points. Talk about the predictions and reasoning. Continue on with Exercises 6-12 and return for sharing and closure. Exercises 13-15 should be assigned to all students.

3 Extend the Activity

Collect the following data from your class—name, height, time to go from class to locker. Create a scatter plot for these data. Ask, "How are these data sets associated and why?"

4 Assess the Activity

Refer to the Weight and Height columns for tenth graders in Table 2 on page 5 or in Table 6 on page 6. Have students complete or answer the following.

a. Prepare a scatter plot for Height versus Weight for sophomores.

b. Is there a positive or negative association? How did you conclude this?

c. Is the association strong or weak? Why?

d. Predict from this data the weight of a 65-inch sophomore.

e. How confident are you in your prediction? Why?

Activity 1-4 How Strong is an Association?

1 Motivate and Focus

Ask students the following questions.

a. What do scatter plots look like when they show strong positive or negative association?

b. How confident can we be in our predictions from the paired data?

c. How can you use a scatter plot to predict?

d. What makes prediction difficult?

2 Share and Summarize

Bring the students back together after Exercise 4 and discuss the positive and negative correlations.

Use Exercises 5 and 6 to help students develop their understanding of correlation.

Exercise 7 could be a group activity used to check how good students' visual estimates were.

Exercise 8 is within everyone's reach while Exercise 9 should be used with students who have a very thorough understanding.

Exercise 10 should be assigned to everyone.

3 Extend the Activity

Copy the lunch menu from school for the last two weeks. Have a student rank the lunches from 1 to 10 with 1 being their first choice. Have another student also rank the lunches from 1 to 10. When the data is gathered, have the students put it on a scatter plot and see if there is any type of association between the two rankings. Have them estimate the correlation coefficient and explain their findings.

4 Assess the Investigation

Have students make scatter plots from Ms. Molar's data in Activity 1-2 on pages 5 and 6 for the variables listed below. Then have them determine whether the association is positive, negative, or no association and estimate the correlation coefficient. Choose the data for the males or for the females.

a. Ninth grade: bench press or leg press versus age

b. Tenth grade: 40-yd dash versus bench press or leg press

c. Eleventh grade: weight versus GPA

d. Twelfth grade: height versus age

On a coordinate axis, have students use 15 data points to produce a scatter plot that has a correlation coefficient of about -0.80.

Have students identify two variables they think may be related. Then have them gather numerical data on the variables and create a scatter plot. Ask them to describe the scatter plot. Give students a value of one variable, and ask them to predict the value of the other.

Have students prepare a table of calories and grams of fat for 15 food items. Then have them create a scatter plot for calories versus grams of fat. Ask them to describe the scatter plot. Then ask, "How many calories would one expect to get from a food that contains 13 grams of fat per serving?"

Line Fitting

Mathematical Overview

In order to use a scatter plot to predict values for one variable on the basis of a value of a related variable, consolidation or summarization of the data is needed. The desired summarization is often achieved by a "summary" line drawn through the middle of the scatter plot. In this investigation, students draw an eye-ball summarizing line using string or a ruler. Comparison of lines drawn by different students suggests some lines seem to better fit the data than others. One indicator of "goodness of fit" is the average closeness of data points to the line. The distances from the points to the line are the "residuals", and their average (mean residual) is used as a measure of how well the line fits the data.

Investigation Outcomes

Students will

- visually estimate summarizing lines for scatter plots;
- use summarizing lines to make predictions;
- describe characteristics of a good summarizing line for a scatter plot;
- determine and use mean residuals to compare scatter plot summarizing lines.

Materials Needed

 graph paper

 string

 ruler

Activity 2-1 Getting to Know Yourself and Others

1 Motivate and Focus

For Olympic sprinters in 1992, what would have been a reasonable time goal for running 100 meters in order to be competitive? How might we consolidate and summarize the data in a scatter plot so that prediction is easier?

2 Share and Summarize

Following group discussion of the first two paragraphs, have students share ideas to ensure group understanding.

Bring students together after Exercise 7. Have a recorder from each group share the characteristics used by the group to decide which line was more accurate in summarizing the data in the table.

Share and summarize again after Exercise 13. Note that narrow ovals clearly suggest where the summarizing line should go. Circular ones provide little help.

Following the pooling of group data toward the end of each project, conduct a whole class discussion of the remaining questions.

Note that as students progress through this unit they will develop systematic methods that will help them to answer the questions more completely.

3 Extend the Activity

Interview a coach or athletic director at your school to find out what kind of data he/she keeps and how he/she uses that information. What types of data are kept on students at your school and how are they used? Seek some paired data for scatter plot representation and analysis.

4 Assess the Activity

Use Exercise 16 as an assessment.

Have students choose two American League team statistics from the tables published in *USA Today* every Tuesday during baseball season that they think might be related to winning percent. Then have them:

a. Prepare scatter plots of winning percent versus variable 1 and versus variable 2.

b. Comment on how good their judgement of association was.

Activity 2-2   How Good is Your Fit?

1 Motivate and Focus

Ask students to predict the women's 1992 high jump winning height by looking at the table. Ask, "On the basis of the data, do you think it went down? went up? stayed the same?"

After students construct the scatter plot, have them predict again. Then have them compare their estimates to the official 1992 high jump winning height data from the Barcelona Olympics (6 ft $7\frac{1}{2}$ in.).

Ask students the following questions.

● How close were your estimates?

● How might we determine "closeness" in different mathematical situations?

2 Share and Summarize

Say, "This activity deals with measuring how closely your summarizing line fits a scatter plot."

Bring students together after Exercise 4 to share and discuss their mean residuals. Discuss differences and how the mean residual is used to estimate the "closeness" or goodness of fit.

Assign Exercise 5 with the stipulation that the groups are to find a line with a "better" fit, in terms of mean residual, than the line drawn in Exercise 2. Emphasize the need to look at the previous line for potential strategies for improvement. Share the results following the group attempts.

Discuss the impact and importance of 1956 and 1960 times on the line of best fit in Exercise 8a. Also discuss whether they should be included in the sample.

3 Extend the Activity

Use *Statistics Workshop* from Sunburst to explore the moveable line feature from the scatter plot menu. Generate three summarizing lines for the women's running high jump data.

A possible group project would be to list the women's records in swimming and in track and field that were broken in the 1992 Olympics. For these events, have students make scatter plots of winning time versus year and compare the predicted value with the actual winning times.

Get a copy of the football data from the athletic department in your school for last season. Have students construct scatter plots for completions versus passing attempts and for penalties versus wins and for others pairs they choose.

Ask students the following questions.

● What patterns do you see?

● What advice might you give the coaching staff?

4 Assess the Investigation

Have students make a scatter plot using data from Table 1 or Table 5 on pages 5 and 6 for height versus 40-yard dash time. Then have them determine the following from the table and the scatter plot.

a. Does the graph have a positive, negative, or no association?

b. Draw a summarizing line, if the data are at all linear.

c. Estimate the goodness of fit by computing the mean residual.

d. Predict the height of a ninth grader who runs the 40-yard dash in 5.8 seconds. Write a sentence expressing your confidence level.

Ask students to identify two variables they think may be related. Then have them gather numerical data on the variables and create a scatter plot and a summarizing line. Given a value of one variable, have students predict the value of the other and comment on their confidence in making the prediction based on the graph and line.

Have students prepare a table of calories and grams of fat for 15 food items. Then have them create a scatter plot for calories versus grams of fat and a summarizing line. Ask students to predict, using the line, the number of calories they would expect to get from a food that contains 9 grams of fat per serving. Have students find foods with approximately 9 grams of fat per serving and determine how closely the calorie count fits the prediction.

Median-Fit Line

● **Mathematical Overview**

This investigation leads students through the process for drawing a median-fit line on a scatter plot that shows a linear association. It becomes the basis for investigations in the remainder of the unit. The process for determining the median-fit line will generate lines that are more closely alike than the visual methods used in the prior investigation. Because of the importance of the median-fit line later, time spent here in helping students develop skills to quickly and accurately find the line for a graph will pay large dividends in future work.

● **Investigation Outcomes**

Students will
● find the median-fit line for a scatter plot;
● use the median-fit line to make predictions.

● **Materials Needed**

 ruler

 tracing paper

 software

 string

 calculator

Activity 3-1 Finding and Using Median-Fit Lines

1 Motivate and Focus

Examine several student-drawn scatter plots and lines from the last investigation to illustrate the variety of summarizing lines that arise from visual approaches.

Ask students the following questions.
● Why do the lines vary?
● What impact do the variations have on predictions?
● What do we need to do to reduce variation?

The lead-in paragraphs identify for students one type of summarizing line that can be done by a computer. Ask students what measure of center should be key to constructing the median-fit line.

2 Share and Summarize

After student groups have worked Exercises 1 and 2, bring the groups together to share patterns found and to discuss the linearity of the data. Point out to students that the vertical scale of the graph in Exercise 2 does not begin at 0. This is common in computer-generated graphs.

After student groups work on Exercises 3-8, share the group results and have groups clarify problems they faced in drawing the median-fit line. Have students refer to the table on page 24 as needed. Develop a working procedure for all. Ask, "Do all median-fit lines closely resemble each other?"

Following small group or individual work on Exercises 10-13, share results and questions again. Be sure each student can construct a median-fit line for a scatter plot.

3 Extend the Activity

Gather anonymous data scores on two tests in a class. Have students construct the scatter plot, median-fit line, and several predictions of what a student could expect on the second test, knowing the score on the first test.

Observe students as they construct the median-fit line and evaluate them on their cooperative group efforts.

4 Assess the Investigation

Allow students to design a small experiment in which they choose two variables they think might be related. Have them collect the data on those variables, construct scatter plots and decide whether or not the data deserves summarization with a median-fit line. Ask for complete and careful student descriptions of all aspects of the project.

Linear Patterns

● Mathematical Overview

This investigation ties together the concept of median-fit line and patterns that are linear. The student will see how linear patterns can be expressed as tables, rules, graphs, and equations. Values which contain negative numbers will let the student extend the graph into other quadrants of the coordinate plane. Some predicting will be done using one- or two-step equations, modeling situations which will be familiar to the students.

● Investigation Outcomes

Students will

● write verbal rules and equations for median-fit lines;

● use the equations of median-fit lines to make predictions;

● extend the median-fit line to all quadrants to make predictions using negative values;

● recognize when data can be represented by a linear model.

● Materials Needed

 graph paper

 ruler

 calculator

Activity 4-1   Representing Linear Patterns

1 Motivate and Focus

Have students look at one of the examples of a median-fit line from Investigation 3. Ask students to predict exact answers from such situations. Start a discussion asking students under what circumstances exact answers could be found.

Look for ideas such as:

● for each value there is only one corresponding value;

● if a line is being used to do the predicting, then prediction is most accurate when all points lie on the line.

Ask students if they can identify a situation that will give paired data that all lie in a line. Tell them they will examine situations such as this in this activity.

2 Share and Summarize

Make sure students have answered Exercise 3 using the proper coordinates. Relate to all the rule developed in Exercise 5a and the equation of Exercise 5d. Bring students together after Exercise 6. Discuss the idea of renting $8\frac{1}{3}$ tapes as the equation used to answer Exercise 6c suggests.

After Exercise 10, bring groups together to examine the graphs in Exercises 2 and 8. You may need to review the concept of correlation. Look for similarities and differences. Seek student ideas for solving Exercise 10. Help them see the use of negatives.

A final sharing and summarizing is appropriate after Exercise 19. Seek student views of relations among rules, graphs and equations.

In Exercise 21, you may need to review that 5% = 0.05. In Exercise 22d, discuss appropriateness of answers. Ask, "Is $326.625 an appropriate answer to this question?"

3 Extend the Activity

Test some of the situations students thought might be linear in the Motivate and Focus section to see if they really are linear. Have students gather data, make a scatter plot, and fit a line.

4 Assess the Activity

Give students the following problem.

Bob, the plumber, charges $30 to come to the house and $18 per hour for his service.

a. Make a table showing Bob's earnings for 0, 1, 2, 3, 4, 5, and 6 hours of labor.

b. Make a scatter plot of the plumber data. Draw the median-fit line.

c. Where does the median-fit line intersect the x-axis? What does this represent?

d. Write a sentence stating a rule that seems to describe the linear association.

e. Express your verbal description as an equation.

Activity 4-2   Linear Equations

1 Motivate and Focus

Discuss how sales personnel are paid and the idea of commission. Set up some simple situations where you will earn a 10% commission on certain sales and find the amount of commission. Then add the idea of getting a salary plus a commission. Notice that this is a two-operation process. Many of our patterns from paired data can be represented with a two-operation equation. These will be evident in the activity.

2 Share and Summarize

Before students begin to work on Exercise 1, you may want to review integer operations. It may not be common practice to use negative commissions, but the situation presented in Exercise 2b does. Bring students together after Exercise 3 to share and discuss their ways of graphing using negative coordinates. Stress the concept of order when plotting points. Make sure students understand how to locate points on a coordinate system by having them locate several points named by ordered pairs and then by writing the ordered pair for each of several points that are plotted on a coordinate system.

Stop to check progress after Exercise 5. Check to see how the groups are working with the two-operation equation necessary there. Take time to assess their understanding of the meaning of the constant and where the coefficient of x came from.

The step graph in Exercise 11c may need some discussion time so students can understand it.

For Exercise 15, point out that generally the age data will produce scatter plots with all points lying on a line. However, if a student is not promoted for a year, the resulting scatter plot will not be points on a straight line.

In Exercise 16a you may need to discuss the case where $x = 0$. Depending upon what points are chosen, the graphical representations in Exercise 16b may be very different.

A brief discussion of the concept of domain may be appropriate for Exercise 17a.

3 Extend the Activity

Draw the graph of a linear equation. From the graph, ask students to name points on the line, describe a pattern relating x and y, and then make an equation describing their findings.

4 Assess the Investigation

Give students the following problem.

Willa Williams, from Walla Walla, Washington, sells widgets. She gets 10% commission on the amount of widget sales and a $200 weekly salary.

a. Make a table of Willa's earnings for the following sales values: $0, $300, $500, $800.

b. Make a scatter plot and median-fit line for the data.

c. Write an equation describing Willa's earnings.

d. Find her earnings on sales of $1200.

e. Explain how you can tell when an equation is linear.

Linear Functions

● Mathematical Overview

The first activity in this investigation stresses the idea that points whose coordinates satisfy a linear function lie on a line. Technology is needed to bring this home visually and independently. Activity 5-2 provides students with the opportunity to investigate the graphical effects of change in m (slope) and b (y-intercept). The investigation concludes with an activity that leads the student toward an algorithm to aid in determining the slope of a line. The entire investigation is an excellent opportunity for using graphing technology in the instructional process.

● Investigation Outcomes

Students will

● represent linear functions with tables, equations, and graphs and relate each to the other;

● graph linear functions described by tables and equations;

● compare the slopes and intercepts of linear functions using equations and graphs;

● determine the slope of a line from its graph, its equation, or two points on the line;

● find the equation of a line given two points on the line.

● Materials Needed

 graph paper   *colored pencils*

graphing calculator or computer function grapher

Activity 5-1   Equations, Tables, and Graphs

1 Motivate and Focus

Discuss hourly wages. Have students who work explain their pay system. If no one works, use the hourly wage of $5.00. Generate a table of values and graph them. Discuss the type of graph formed. Ask students to name other situations that could be represented by such a graph.

2 Share and Summarize

Bring students together after Exercise 2. Discuss methods used to find values for a table when given x or when given y. Be sure that all students can use the technology to graph a function, set ranges, plot points, and use the TRACE command. Setting the range to square coordinates using that zoom feature of the graphing calculator may be helpful. After Exercise 6, you may want to discuss relative pixel width. After Exercise 9, discuss the graph of a linear function and guide students to see that it is always a line. Exercises 10-12 are better done by paper and pencil since some graphing utilities require equations in the form $y = mx + b$.

3 Extend **the Activity**	Have students graph the equation that relates the number of blooms on a cactus (y) to the number of days of sun it gets in a month (x). The equation is $y = 7x - 1$.
4 Assess **the Activity**	Informal assessment can take place in the classroom by providing other equations for students to work on individually. Give students the following problem. A classmate says an equation like $y = -2x + 1$ has a line as its graph.

a. What does it mean to say "an equation like $y = -2x + 1$"?

b. Is the classmate correct? Explain why or why not.

Activity 5-2 Visualizing Slopes and Intercepts

1 Motivate **and Focus**	Ask students the following questions. ● How are the graphs of $y = 3x$ and $y = 5x$ similar? ● How are they different? How are they similar and different from $y = 2x$? Tell students to use Exercises 1 and 2 to investigate this issue.
2 Share and **Summarize**	Bring students together after Exercise 1 to share and discuss their findings and thoughts. Work out any disagreements and gain a consensus. Check for group understanding by assigning of Exercise 2, either individually or for group processing. In Exercise 2, have students graph $y = x$ and $y = 2x$ first. Share and summarize again after Exercise 5 to be sure that the idea of y-intercept is developed.
3 Extend **the Activity**	Have students sketch graphs of linear equations with given slope and intercepts. Make some positive, some negative, and some fractional.
4 Assess **the Activity**	Give students the equation $y = 3x - 2$, and have them tell everything they can about its graph and why. Give students the equation $y = -2x + 4$, and have them write a different equation whose graph has:

a. the same slope.

b. the same y-intercept.

Activity 5-3 Calculating Slopes and Finding Equations

1 Motivate and Focus

Ask students the following questions.

- What is the y-intercept of the graph of $y = 5x - 2$?
- What is its slope?

Say, "Given an equation, you can determine the y-intercept and the slope by inspection. Suppose you are given a line in the coordinate plane, could you determine its slope? This activity helps you answer that question."

2 Share and Summarize

You may wish to do Exercise 1 as a whole-class discussion.

After all groups have completed Exercise 4, bring the class together to share conjectures and arguments. Do not continue until good generalizations are agreed to by all. If agreement cannot be reached, have all groups try another linear function such as $y = 0.75x - 1$.

You may want to discuss the slope of a vertical line (no slope) after students answer Exercise 8b.

In Exercise 9b, check to make sure students are progressing with the task. Some students may choose to use graphing calculators.

After all groups have completed Exercise 9, bring the class together to share procedures and to analyze each. Bring the group to an agreement regarding an appropriate procedure such as:

- Use the x- and y-coordinates of two points to find the slope of the line.
- Substitute the slope and the coordinates of one point for m, x, and y to calculate b.
- Write the final equation by substituting the calculated values for m and b in the equation $y = mx + b$.

3 Extend the Activity

Ask students, "How are the graphs of the lines $y = mx + b$ and $y = -\frac{1}{m}x + b$ related to each other?" **The lines are perpendicular to each other and intersect at (0, b).**

4 Assess the Investigation

Have students graph their answers for Exercise 10 on a graphing utility and use the trace command to find the points listed in the exercise.

Have the students complete the activity given above and write a short summary telling how well they did and, if they made errors, what caused the errors and how they overcame them.

Using Linear Functions

Mathematical Overview

The purpose of this investigation is to make predictions using the equation of a median-fit line for a set of data points. Students will use the equation to solve for x if y is known, or to solve for y if x is known.

The investigation offers two different strategies for solving for the unknown variable. The first uses the graph of the scatter plot and median-fit line to determine an unknown value, usually finding the y value for a given value of x. The second uses cups and counters to develop a manipulative technique for solving algebraic equations of the form $y = mx + b$ for x when given y.

Investigation Outcomes

Students will answer questions about situations modeled by linear functions of the form $y = ax + b$ by:

● finding and interpreting a value for y given a value for x, and

● finding a value for x given a value for y.

Materials Needed

 tape measure

 calculator

 graph paper

 software

 cups and counters (Homemade counters can be made of colored paper or laminated material.)

Activity 6-1 Equations for Fitted Lines

1 Motivate and Focus

Have students give their height in inches and their shoe size. Record these on a chart. Then graph them using height as the independent variable and the shoe size as a dependent variable. Ask students, "Is the relationship between the height and the shoe size a linear function? Can you predict anything from the graph?"

2 Share and Summarize

Remind students that stride length is represented on the y-axis and height is represented on the x-axis. Students may use technology.

Bring the class back together after they have recorded the required data and completed work on Exercises 1-6.

In Exercise 7a, have the students express the times in seconds.

For Exercise 9a, you may want to have students discuss methods for measuring the heights of the bounces.

Discuss the accuracy of using the graph of the line versus using the equation of the line when predicting values of y for a given value of x.

3 Extend the Activity

Give students the following problem.

The varsity club wants to sell bumper stickers. Company A will sell them the stickers for 22¢ each. Company B charges a $50 flat rate plus 11¢ a sticker.

a. Have students draw a graph to represent both plans.

b. Discuss when one company would be cheaper than the other, and when they would be equal.

4 Assess the Activity

Use the three data tables in Exercise 7 as contexts. The first table could be completed in class as a group activity and the following ones assigned as homework or as an outside of the classroom project. One or more of these data sets may be the basis for a portfolio project.

Activity 6-2 Manipulating Linear Equations

1 Motivate and Focus

Write the equations $27x + 43 = 97$ and $2x + 2 = 6$ on the chalkboard. Ask students how these two equations illustrate the "solve a simpler problem strategy."

2 Share and Summarize

After you model the equation in Exercise 5, tell students the goal is to get the cups by themselves on one side of the mat using the rules stated below.

● A **zero-pair** is formed by pairing one positive counter with one negative counter.

● You can remove or add the same number of counters to each side of the equation mat.

● You can remove or add zero-pairs to either side of the equation mat without changing the equation.

Bring students together after they have completed Exercise 6 of this activity. Share solution procedures and difficulties. Move toward a common procedure that is understood and agreed to by all. Practice the procedure using the material in Exercise 7. This exercise provides review of operations with decimals and fractions. If students have no difficulties, then all parts of Exercise 7 need not be completed.

Exercie 9, 10, and 12 provide a great opportunity for using graphs to check answers.

3 Extend the Activity

Ask students how the equation $2x + 3 = x + 7$ is different from the equations studied so far. Then have students try to model and solve the equation.

4 Assess the Investigation

Using the following (x,y) data set: $\{(1, 8), (4, 14), (10, 26), (0, 6), (6, 18)\}$, have students:

a. Plot the points and make a scatter plot.

b. Construct the median-fit line for the plot.

c. Find the equation of the line.

d. Use the equation of the line to find the value for y when $x = 62$.

e. Use the equation of the line to find the value for when $y = 200$.

Refer students to the baseball data used to assess Activity 2-1 on page T16, and then have them:

a. Make a table showing the data on average home attendance, average away attendance, and winning percent for all the American League teams.

b. Contruct scatter plots for average home attendance versus winning percent, average away attendance versus winning percent, and average attendance versus winning percent.

c. Find the equations of the median-fit lines for those scatter plots that are linear. Decribe why the other plots are not summarized with a line.

d. Predict attendance, home, away, and total, using the winning percents of 0.45, 056, and 0.67.

e. Predict winning percent using average total attendance of 25,000, of 28,000, and of 32,000.

Teacher's Answer Key

Page 2 **1a.** Bench press is the amount of weight on one bar bel that is lifted using both arms while lying on a bench. GPA is the grade point average using a 4-point scale in this case. Leg press is the amount of weight lifted using the legs while sitting on a bench.

2a.

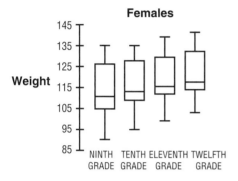

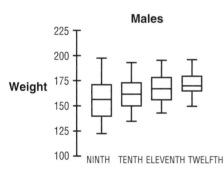

3a.

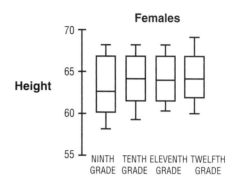

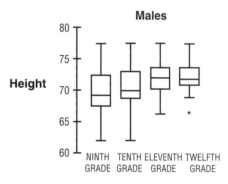

Page 3 **2b.** For both genders, the weights varied the most in ninth grade. The median weights increased yearly and the range of weights became smaller. **2c.** Typically students use the median because it is easily accessible from the box-and-whisker plots. Answers will vary.

3b. Sample answer: For females, the distribution varied the most in 9th grade; in tenth grade the median height increased and remained stable. For males, the median height increased until eleventh grade, when it stabilized; the distributions varied less each year. **3c.** Answers will vary. If students again use the median, 64 inches would be typical in grades 10, 11, and 12 and 62.5 inches in 9th grade. **4a.** Answers may vary because of the comparison using different units. Encourage students to support their choices with reasoning. Generally, the smaller the interquartile range and length of whiskers, the less variation in the data. **6b.** The median leg-press weight increases by 12.5 pounds between ninth and tenth grades and then by 7.5 pounds every year thereafter.

Page 4 **8.** Sample answer: Comparing the mean and median times of the six youngest females with those of the six oldest females suggests that the older females are faster. However, the five fastest runners are between 165 months and 179 months old which suggests no age constraints.

Page 9 **7.** Yes, they are associated. No, the association is not positive. As the height increases on the *x*-axis, the fingertip to ceiling distance decreases on the *y*-axis.   **8.** Sample answer: The points of the scatter plot fall as the values on the horizontal axis increase. An increase in one variable is accompanied by a decrease in the other.

Page 10.   18b.

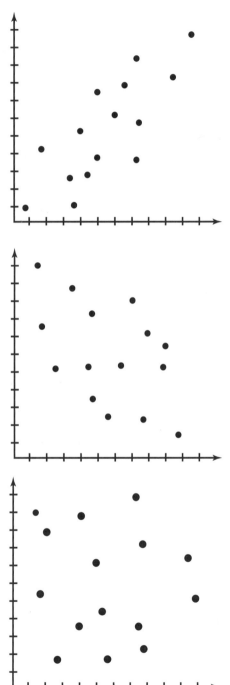

Page 11   1a.

NINTH GRADE LEG PRESS VERSUS BODY WEIGHT

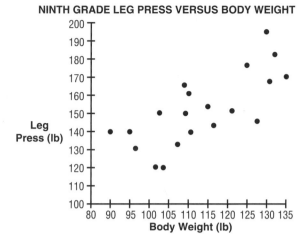

NINTH GRADE BENCH PRESS VERSUS BODY WEIGHT

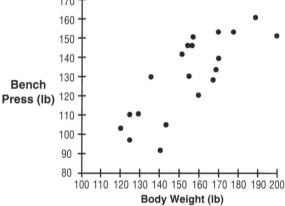

1c. Moderate; The points do not appear to lie on a line, yet do follow an uphill pattern. (You may wish to ask students to illustrate strong, moderate, and weak associations.)
2. There are no clusters for the females and possibly 2 clusters for the males. Each cluster contains 3 data points in which the weight of the male is consistent and the corresponding bench-press weight is within 10 pounds of the others.   **4.** Sample answer: A bench press of 170 might imply a body weight of about 205 pounds. A leg press of 145 might imply a body weight of about 108 pounds. Draw a horizontal line (*either* $y = 170$, *or* $y = 145$) and estimate where a vertical line might intersect it, keeping in mind the given data.

6.

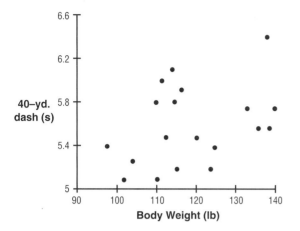

ELEVENTH GRADE FEMALES

40–yd. dash (s) vs. Body Weight (lb)

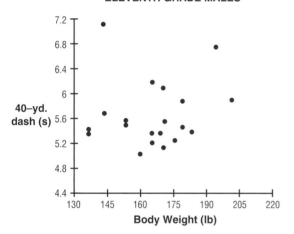

ELEVENTH GRADE MALES

40–yd. dash (s) vs. Body Weight (lb)

Page 12. **7.** Yes, there is a weak, negative association. Since body weight and 40-yd dash times are postively associated, body weight and speed are negatively associated. Time to run the 40-yd dash and speed of running the 40-yd dash are inversely proportional. **8.** There are no clusters in either scatter plot. The females had one possible outlier with respect to time only. The males had two possible outliers, one with respect to weight only, the other with respect to weight and time. **9.** No. Because of the weak association of the data, the prediction would not be reliable. Also, in each case, the corresponding data points exhibit a large range.

10b.

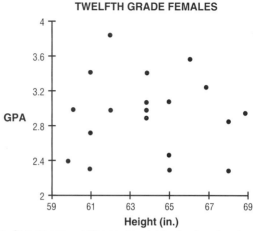

TWELFTH GRADE FEMALES

GPA vs. Height (in.)

10c. Sample answer: The scatter plot for the females shows a random scatter suggesting little to no association. The scatter plot for the males shows a fairly random scatter, however one might notice a slight negative association. **10d.** Answers will vary. Because of the small association between variables, confidence in predictions will be low.

11a.

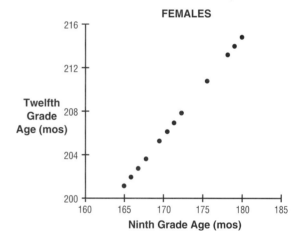

FEMALES

Twelfth Grade Age (mos) vs. Ninth Grade Age (mos)

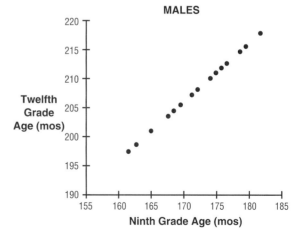

MALES

Twelfth Grade Age (mos) vs. Ninth Grade Age (mos)

11b. Positively; As the age of the ninth graders increase, clearly their corresponding ages in twelfth grade will increase as well.

Page 13   14. Sample answer: If the association is positive, the line rises from left to right. If the association is negative, the line falls from left to right.   **15.** Sample answer: If a line can be drawn through the set of points with all points relatively close to or on the line, a strong association is present. The larger the relative total distance from the points to the line is, the weaker the association.

Page 14   2b. the middle scatter plot because the loop containing these points would be wider than the loops of the other scatter plots   **3.** The scatter plot on the left and in the middle both show a weak correlation. The first shows weak negative correlation, and the second shows weak positive correlation.   **4.** Correlations between 0 and 21 represent negative associations; the stronger the negative association the closer the number is to 21.

5a.

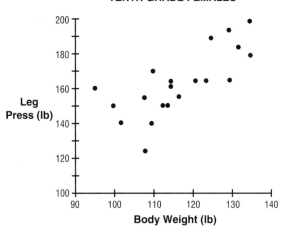

TENTH GRADE FEMALES

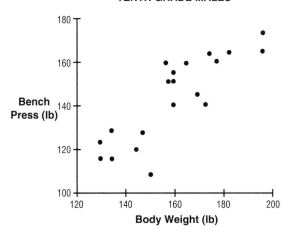

TENTH GRADE MALES

5b.

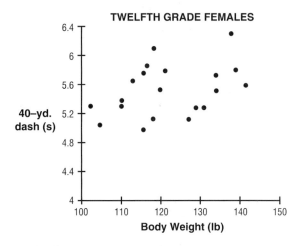

TWELFTH GRADE FEMALES

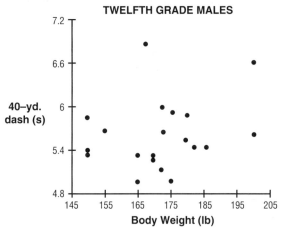

TWELFTH GRADE MALES

5c.

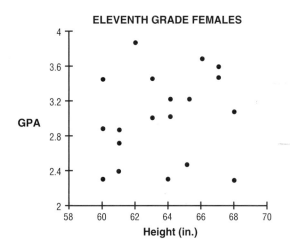

ELEVENTH GRADE FEMALES

5c. (cont'd)

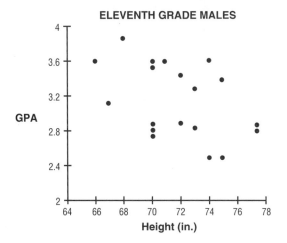

ELEVENTH GRADE MALES

5d.

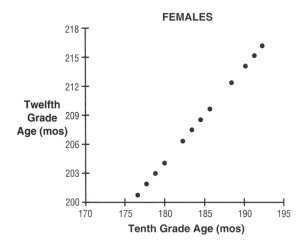

FEMALES

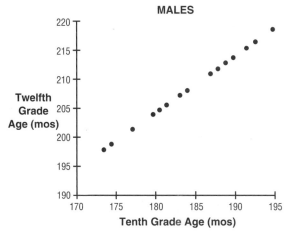

MALES

Page 15.   7. a. Females: 0.75, Males: 0.85; **b.** Females: 0.33, Males: 0.23; **c.** Females: 0.14, Males: 0.45; **d.** Females: 1.00, Males: 1.00

Page 17   2a.

Data Insights: Scatter Plot

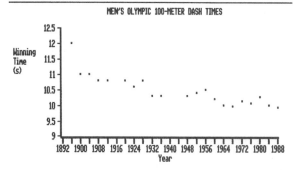

MEN'S OLYMPIC 100-METER DASH TIMES

The horizontal axis represents year of race. The vertical axis represents winning time.   **3d.** Since hundredths of a second were not added until 1972, it is unclear whether the fastest time was 9.9 in 1968 or 9.92 in 1988. Students might investigate the 1992 time.

4b.

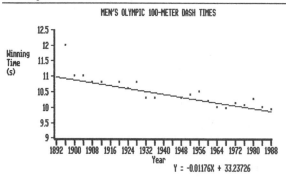
Data Insights: Scatter Plot

MEN'S OLYMPIC 100-METER DASH TIMES

Y = -0.01176X + 33.23726

4d. Answers should be close. Estimates were made easily with the line. A line gives a visual representation for estimation.

Page 18   10a.

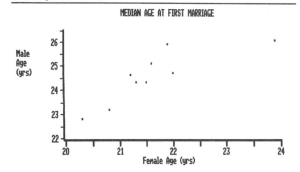
Data Insights: Scatter Plot

MEDIAN AGE AT FIRST MARRIAGE

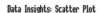
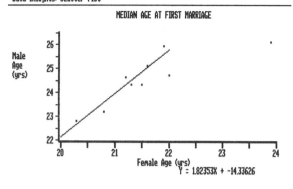

MEDIAN AGE AT FIRST MARRIAGE

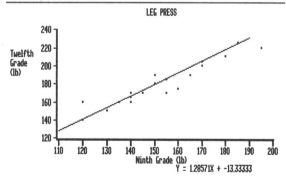

12. No; Since ages generally declined between 1900 and 1970 and have been increasing since, a straight line is not appropriate to model the situation.

Page 20 **14a.** Sample answer: The data seem to have a negative association so a line might be a fine model for this data. One possible problem would be predicting future times based on the line because when the line crosses the *x*-axis the times would be 0 or negative, which is impossible.
14b. Sample answer: wavy curves or curves that follow a decreasing pattern and then level out at some point
15b.

Data Insights: Scatter Plot

LEG PRESS

Twelfth Grade (lb)

Ninth Grade (lb)
Y = 1.28571X + -13.33333

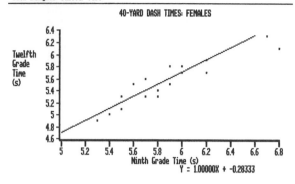
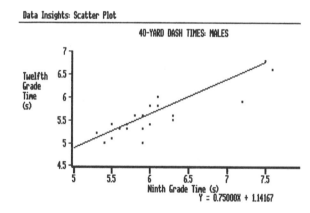

16a. Begin by looking for an association between the variables in question. If a strong association exists, try to draw a line that minimizes the total distance from all points to the line.

Page 22 2a.

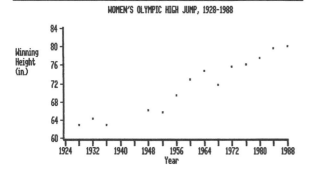

Winning height is the *y*-axis label, and Year is the *x*-axis label.
3d. Answers may vary. Actual conversions will depend upon the size of the scatter plot printout. Distances should represent points farthest away from and closest to the line.
6c. Answers may vary. Students should be most confident in the prediction based on the line with the smallest mean residual.

Page 23 8d. Answers may vary. It appears that the winning times are decreasing less rapidly with time and depending on interpretation of data, confidence may or may not be good.

Page 25 2. Sample Answer: No; See students' work.; A summarizing line extending from 14 on the *y*-axis to 1985 on the *x*-axis will have 8 points above the line and 8 points below the line.

Page 27 5b. Sample answer: Approximately 10.3 deaths per thousand citizens; About 0.7 deaths per thousand citizens.
7a. Sample answer: For the first half of the century, the median-fit line seems appropriate, but the death rate appears to be leveling out during the last half of the century, and the line is not very appropriate.
9.

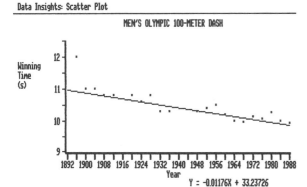

Page 27 10a.

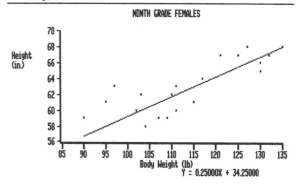

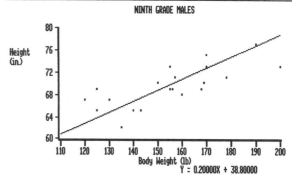

10c. Sample answer: About 135 pounds for the male, and about 127 pounds for the female.

Page 28 11a.

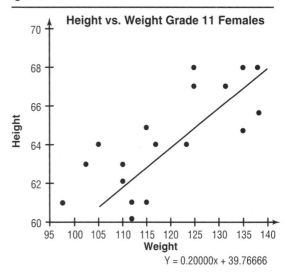

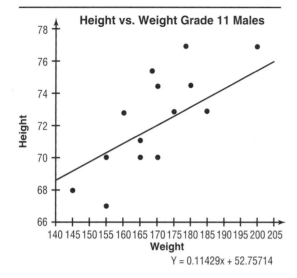

Height vs. Weight Grade 11 Males

$Y = 0.11429x + 52.75714$

12a.

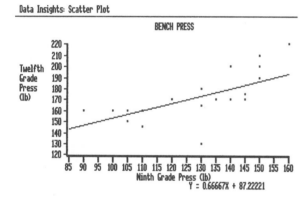

Data Insights: Scatter Plot

BENCH PRESS

$Y = 0.66667X + 87.22221$

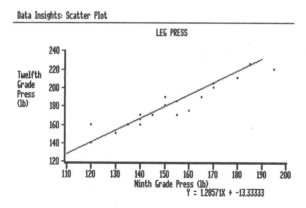

Data Insights: Scatter Plot

LEG PRESS

$Y = 1.28571X + -13.33333$

12b. Sample answer: Bench press goals: 175 pounds, 200 pounds, 220 pounds; leg press goals: 155 pounds, 190 pounds, 215 pounds.. **12c.** Bench-press means: 9th grade is 129.5 pounds, 12th grade is 173.75 pounds; Leg-press means: 9th grade is 152.5 pounds, 12th grade is 180.25 pounds; these points are extremely close if not on the line.

13a.

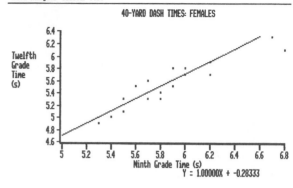

Data Insights: Scatter Plot

40-YARD DASH TIMES: FEMALES

$Y = 1.00000X + -0.28333$

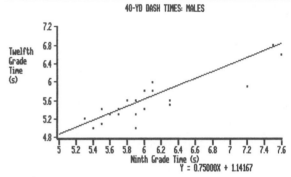

Data Insights: Scatter Plot

40-YD DASH TIMES: MALES

$Y = 0.75000X + 1.14167$

13b. Sample answers: Male goals: 5.95 seconds, 5.4 seconds, 5.55 seconds; Female goals: 6.2 seconds, 5.5 seconds, 5.65 seconds.
14a.

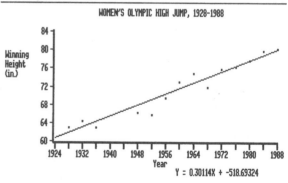

Data Insights: Scatter Plot

WOMEN'S OLYMPIC HIGH JUMP, 1928-1988

$Y = 0.30114X + -518.69324$

Page 30   15d. They are difficult to interpret because of the wide range of values for each variable. In all three cases, outliers appear to exist. The outliers represent data from the Atlantic, Pacific, and Indian Oceans.   **15e.** Relationships are more easily observed without outliers.

15e. (cont'd)

Data Insights: Scatter Plot

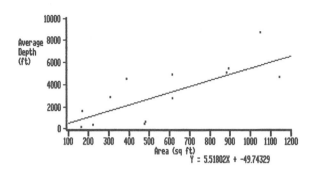

$Y = 5.51802X + -49.74329$

Data Insights: Scatter Plot

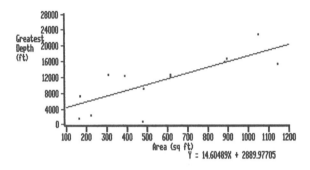

$Y = 14.60489X + 2889.97705$

Data Insights: Scatter Plot

SEVENTEEN LARGEST OCEANS & SEAS

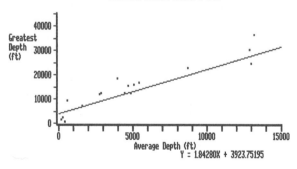

$Y = 1.84280X + 3923.75195$

f. Answers may vary. Using Data Insights software the following equations represent median-fit lines:
Average Depth vs. Area, $y = 5.52x - 50.77$;
Greatest Depth vs. Area, $y = 14.61x + 2887.27$;
Greatest Depth vs. Average Depth, $y = 1.84x + 3923.73$ (The first two graphs are excluding the first four oceans; the third graph includes all data.)   **16a.** Answers will vary. One possibility could be to divide data into quartiles, calculate each quartile's mean, then connect the first and fourth quartile mean points and slide the line parallel to the middle of the 2nd and 3rd points. Another possibility is to use the median-fit procedure with the means rather than medians.

Page 31   2a.

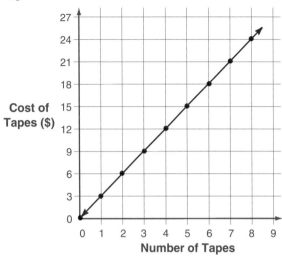

Page 32   4c. No. Even if each tape was only $1\frac{1}{2}$ hours long, she would need 75 hours to watch the tapes. There are only 24 hours in 1 day.   **5a.** The total cost of the rental, y, equals $3 times the number of tapes rented, x.

Page 33
8a.

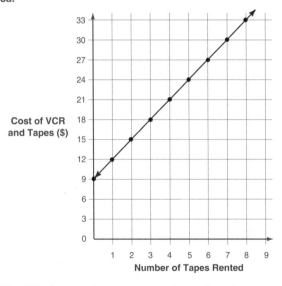

9b.   The first coordinate represents the number of tapes rented, in this case 0. The second coordinate represents $9 for the cost of the VCR and $3 for each tape rented.   **11c.** No. If each tape was only $1\frac{1}{2}$ hours long, Cedric would need at least 52.5 hours to watch all the tapes and there are only 24 hours available for viewing.

Page 34
16a.

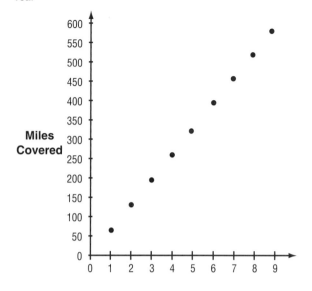

Miles Covered

16b.

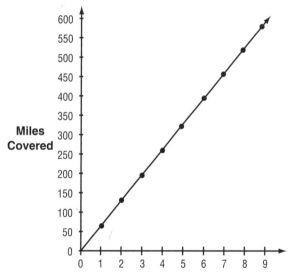

Miles Covered

Page 35
20a. Sample Answer: Use a ruler to mark where $y = 300$ intersects the median-fit line and again use the ruler to determine where the vertical line through the intersection meets the x-axis. It takes approximately $4\frac{1}{2}$ hours to drive 300 miles.

Page 35
21b.

Total Sales ($)	0	1,000	2,000	3,000	4,000	5,000	6,000	7,000	8,000	9,000	10,000
Weekly Earnings ($)	300	350	400	450	500	550	600	650	700	750	800

Page 35   21c.

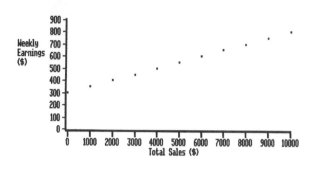

21d.

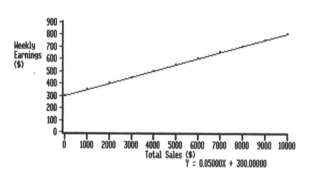

The line intersects the y-axis at (0,300) and the x-axis at (26000,0).

23a.

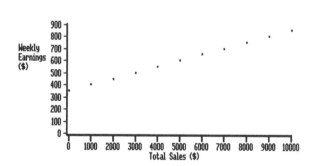

23b. Graphs should be very similar, lines containing the points should be parallel. The *y*-intercept is 350, and the *x*-intercept is -7,000.

Page 36   3a.

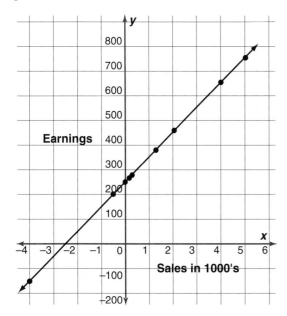

3b. The *x*-axis represents sales in $1000 increments. The *y*-axis represents earnings in $100 increment Points to the left of 0 on the horizontal axis are negative. Points below 0 on the vertical axis are negative.

Page 37 5a. For a given weekly sales value *x*, multiply it by 0.1 and add 250 to find John's total earnings.

6b.

Hours	0	1	2	3	4	5	6	7
Total Cost ($)	25	55	85	115	145	175	205	235

7c.

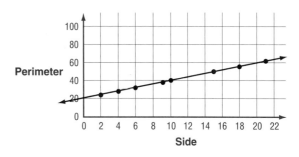

7e. $s = -1$; No, it is not reasonable to represent the length of a side with a negative number.

Page 38   8c.

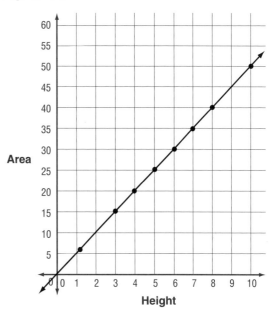

8d. No. The idea of a negative number representing a length of a side does not make sense.

9a.

price: *p*	10	15	20	25	30	40	*p*
Cost: *C*	10.50	15.75	21.00	26.25	31.50	42.00	1.05p

10c.

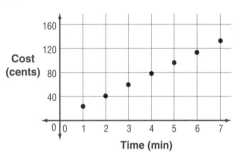

11c.

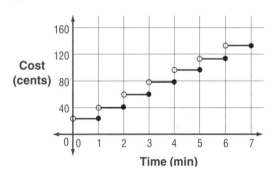

12a.

xy	
1	9
2	8
3	7
4	6
5	5
6	4
7	3
8	2
9	1

13a.

x	y	
1	−9	
2	−8	
3	−7	
4	−6	
5	−5	
6	−4	
7	−3	
8	−2	
9	−1	

Page 39   14b. No. The lines intersect at (10, 0). One line rises from left to right, and the other line falls from left to right. The amount of slant is the same only in different directions.
15a. Only the age data, when compared, will produce scatter plots with all points lying on a line. For example, ninth grade boys versus eleventh grade boys.

16a.

xy	
1	24
2	12
3	8
4	6
5	4.8
6	4
8	3

Page 40
2a. 1,500 represents her monthly salary. 50 represents the amount she is paid for each car sold.
2c.

Data Insights: Scatter Plot
_____

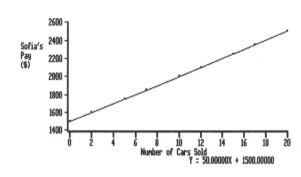

Page 41
3a.

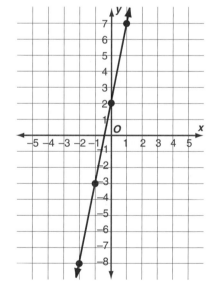

3b.

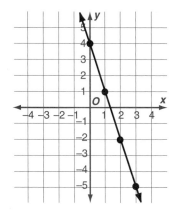

3c.

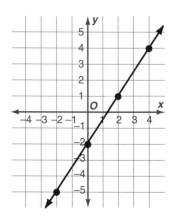

Page 42

7a. Using the TI-81, trace along the graph to point with
x-coordinate of interest. Read off the corresponding y value.
Alternativelyby hand, begin at the x value of interest on the
x-axis. Move vertically to the line $y = 2x + 1$. From that
intersection, move horizontally to the y-axis. This final
intersection indicates the y value in question.

8a.

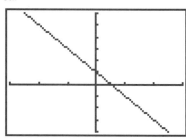

8b.

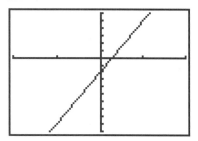

8c.

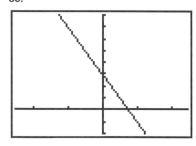

8d.

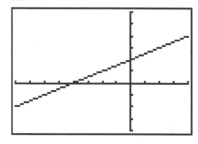

8e.

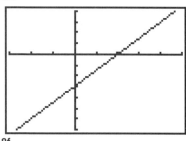

8f.

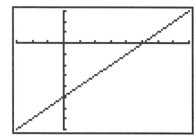

11b. Sample answer: Construct a table of values that satisfy
the equation and plot the corresponding points on a coordinate
grid. The graph is a straight line.

Page 43
1a. If using technology, make sure students can graph several functions on the same coordinate system. Make sure fractional values are used.

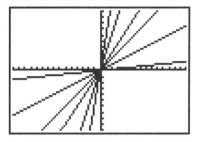

3a.

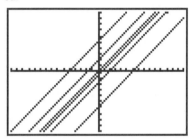

3b.

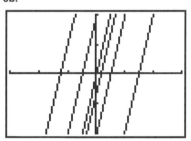

3c.

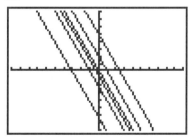

Page 44 **4.** Student descriptions may vary depending upon individual background and interpretation
a. The graph rises steeply from left to right and crosses the y-axis at (0, -6). **b.** The graph falls moderately from left to right and crosses the y-axis at (0, 0.4). **c.** The graph rises gradually from left to right and crosses the y-axis at (0, 0.2). **d.** The graph falls sharply from left to right and crosses the y-axis at (0, 28)

4a-d.

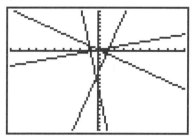

5a. m determines whether the line rises or falls from left to right and also the steepness of the incline or decline.
6c.

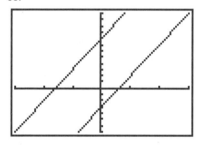

6e.

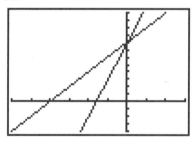

7b-d.

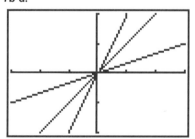

Page 45 **8b.** The x-intercept is the quotient of the constant term and the coefficient of x. Similarly, the y-intercept is the quotient of the constant term and the coefficient of y.
9b. (0, 4) and (3, 0); The x-intercept is the reciprocal coefficient of the x term. Similarly, the y-intercept is the reciprocal coefficient of the y term.

Page 46 **1d.** Yes. The slope is the same as the difference in y-coordinates divided by the difference in x-coordinates.
2b. The y-coordinate differences are 3 times the corresponding x-coordinate differences. **2d.** Yes. The slope is the same as the difference in y-coordinates divided by the differences in x-coordinates.

Page 47 **4a.** Compute the difference for the *y*-coordinates of two points and divide by the difference of the corresponding *x*-coordinates. **7.** Agree. The calculation of the slope is not dependent on which points on the line are chosen

Page 48 **14.** The slope of the road will be $\frac{3}{100}$.

This means that the elevation of the road will change 3 feet vertically for each 100 feet of horizontal road.
15. Sample answer: Suppose (c, d) and (e, f) are two points on the line. Then $d = mc + b$ and $f = me + b$. Subtracting $d - f = mc + b - (me + b) = mc - me = m(c - e)$
and $m = \frac{d-f}{c-e}$ which is the slope of the line.

16c. There are infinitely many lines with slopes between 0 and 1 and a *y*-intercept of 2. **16d.** There are infinitely many lines with slopes greater than -1 and a *y*-intercept of 0.

Page 50
7a.

Data Insights: Scatter Plot

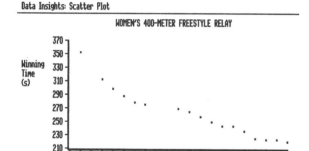

WOMEN'S 400-METER FREESTYLE RELAY

Data Insights: Scatter Plot

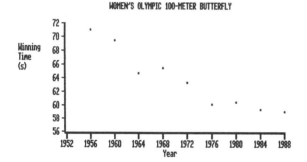

WOMEN'S OLYMPIC 100-METER BUTTERFLY

Data Insights: Scatter Plot

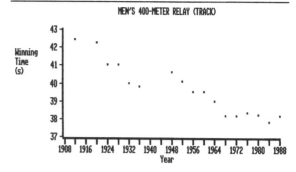

MEN'S 400-METER RELAY (TRACK)

Page 51 **7b.** Answers may vary. The following summarizing lines represent the median-fit line using Data Insights: Women's 400-meter freestyle relay, $y = 1.33x + 2849.16$; Women's 100-meter burtterfly, $y = -0.43x + 905.421$; Men's 400-meter relay, $y = 0.05x + 144.93$ (Equations with less than three decimal places will produce seemingly nonreliable data. You may wish to discuss the implications with students.
7c. Answers may vary. The following 1992 predictions are based on the equations above and are: Womens 400-meter freestyle, 206.81 seconds; Womens 100-meter butterfly, 55.50 seconds; Men's 400-meter relay, 37.48 seconds.

Page 52 **8a.** Sample answers: Women's 400-meter freestyle relay, -1.45; Women's 100-meter Butterfly, -0.386; Men's 400-meter relay, -0.056. **8c.** Womens 400-meter Freestyle, 2134; Womens 100-meter Butterfly, 2137; Mens 400-meter relay, 2673; The x-intercept represents the year in which the winning time will be 0 seconds. It does not make sense to have a winning time of 0 seconds. **9c.** Yes. The height of the second bounce can be found using the height of the first bounce at the x-coordinate representing the original height and finding the corresponding y-coordinate on the summarizing line.
10b. males, y = 0.74999x + 1.04175; females, y = x + -0.18333; The slopes are the same, but the y-intercepts of the new equations are decreased by 0.1. **10c.** males, y = 0.74999x + 1.34175; females, y = x + -0.48333; The slopes are the same, but the y-intercepts of the new equations are 0.2 more than the original line.

Page 53 **10d.** A change of a constant, *k*, to the variable graphed on the vertical axis will result in a change in the *y*-intercept of the representative linear equation by the same amount. **11c.** The summarizing line will be shifted left c units if c is positive and right c units if c is negative. The slopes of the lines will remain the same, but the y-intercept will change by mc.
12a. Depending on *c*, the effect will be a parallel line *c* units left or right. If *c* > 0, the line will appear to be below the original. Students should be able to verify their conjecture with graphs.
12b. Depending on *c*, the effect will be a parallel line *c* units right or left. If *c* > 0 the line will appear to be below the original, and if *c* < 0, the line will appear to be above the original. Students should be able to verify their conjecture with graphs.

Page 54
1a.

1b.

2.

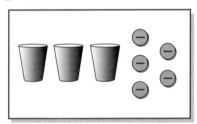

Page 56
6b. Answers may vary. If they do not produce the same value for *x*, they cannot both be correct.
8a.

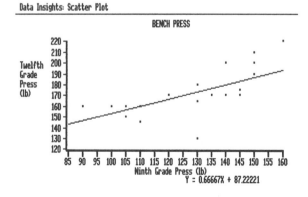

Data Insights: Scatter Plot

BENCH PRESS

Twelfth Grade Press (lb)

Ninth Grade Press (lb)
Y = 0.66667X + 87.22221

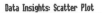
Data Insights: Scatter Plot

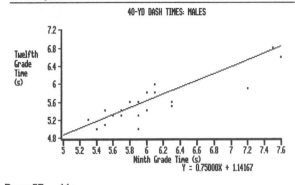

40-YD DASH TIMES: MALES

Twelfth Grade Time (s)

Ninth Grade Time (s)
Y = 0.75000X + 1.14167

Page 57   11a.

Data Insights: Scatter Plot

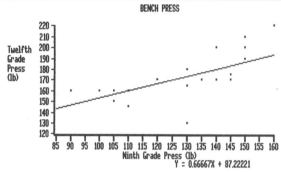

BENCH PRESS

Twelfth Grade Press (lb)

Ninth Grade Press (lb)
Y = 0.66667X + 87.22221

Data Insights: Scatter Plot

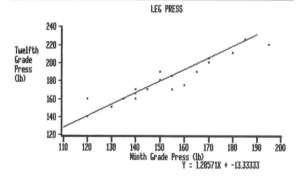

LEG PRESS

Twelfth Grade Press (lb)

Ninth Grade Press (lb)
Y = 1.28571X + -13.33333

Page 61 **1a.**

X	Y1	
0	20	
5	195	
10	370	
15	545	
20	720	
25	895	
30	1070	

X = 15

2a.

X	Y1	
4	0	
4.5	4	
5	16	
5.5	36	
6	64	
6.5	100	
7	144	

X = 7

3. The family of graphs is lines having a *y*-intercept of (0, 4)

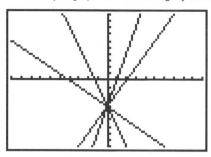

Page 64

1. The family of graphs is lines having a slope of 3.

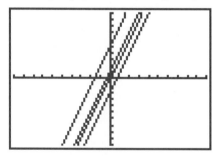

2. The family of graphs is lines that are parallel to the *x*-axis.

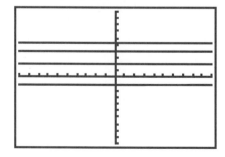

Teacher Notes

Teacher Notes

Teacher Notes

Predicting from Data

An Alternative Unit
for Representing and
Analyzing Two-Variable Data

GLENCOE
Mathematics Replacement Units

GLENCOE
McGraw-Hill

New York, New York
Columbus, Ohio
Mission Hills, California
Peoria, Illinois

Printed in the United States of America.

Send all inquiries to:
Glencoe/McGraw-Hill
936 Eastwind Drive
Westerville, Ohio 43081

ISBN: 0-02-824208-4 (Student Edition)
ISBN: 0-02-824209-2 (Teacher's Annotated Edition)

1 2 3 4 5 6 7 8 9 10 VH/LH-P 03 02 01 00 99 98 97 96 95 94

NEW DIRECTIONS IN THE MATHEMATICS CURRICULUM

Predicting from Data is a replacement unit developed to provide an alternative to the traditional method of presentation of selected topics in Pre-Algebra, Algebra 1, Geometry, and Algebra 2.

The NCTM Board of Directors' Statement on Algebra says,
> "Making algebra count for everyone will take sustained commitment, time and resources on the part of every school district. As a start, it is recommended that local districts–...
> 3. experiment with replacement units specifically designed to make algebra accessible to a broader student population."
> (May, 1994 *NCTM News Bulletin*.)

This unit uses data analysis as a context to introduce and connect broadly useful ideas in statistics and algebra. It is organized around multi-day lessons called investigations. Each investigation consists of several related activities designed to be completed by students working together in cooperative groups. The focus of the unit is on the development of mathematical thinking and communication. Students should have access to computers with statistical software and/or calculators capable of producing graphs and lines of best fit.

About the Authors

Christian R. Hirsch is a Professor of Mathematics and Mathematics Education at Western Michigan University, Kalamazoo, Michigan. He received his Ph.D. degree in mathematics education from the University of Iowa. He has had extensive high school and college level mathematics teaching experience. Dr. Hirsch was a member of the NCTM's Commission on Standards for School Mathematics and chairman of its Working Group on Curriculum for Grades 9-12. He is the author of numerous articles in mathematics education journals and is the editor of several NCTM publications, including the *Curriculum and Evaluation Standards for School Mathematics Addenda Series, Grades 9-12*. Dr. Hirsch has served as president of the Michigan Council of Teachers of Mathematics and on the Board of Directors of the School Science and Mathematics Association. He is currently a member of the NCTM Board of Directors.

Arthur F. Coxford is a Professor of Mathematics Education and former Chairman of the Teacher Education Program at the University of Michigan, Ann Arbor, Michigan. He received his Ph.D. in mathematics education from the University of Michigan. He has been involved in mathematics education for over 30 years. Dr. Coxford is active in numerous professional organizations such as the National Council of Teachers of Mathematics, for which he was the editor of the 1988 NCTM Yearbook, *The Ideas of Algebra, K-12*. He was also the general editor of the 1993 and 1994 NCTM Yearbooks. Dr. Coxford has served as the president of the Michigan Council of Teachers of Mathematics and as the president of the School Science and Mathematics Association. He is presently the general editor of the 1995 NCTM Yearbook.

CONSULTANTS

Each of the Consultants read all five investigations. They gave suggestions for improving the Student Edition and the Teaching Suggestions and Strategies in the Teacher's Annotated Edition.

Richie Berman, Ph.D.
Teacher Education Program
University of California
Santa Barbara, California

Linda Bowers
Mathematics Teacher
Alcorn Central High School
Glen, Mississippi

William Collins
Mathematics Teacher
James Lick High School
San Jose, California

David D. Molina, Ph.D.
E. Glenadine Gibb Fellow in
  Mathematics Education &
  Assistant Professor
The University of Texas at Austin
Austin, Texas

Louise Petermann
Mathematics Curriculum Coordinator
Anchorage School District
Anchorage, Alaska

Dianne Pors
Mathematics Curriculum Coordinator
East Side Union High School
San Jose, California

Javier Solerzano
Mathematics Teacher
South El Monte High School
South El Monte, California

Table of Contents

Making Mathematics Accessible to All: First-Year Pilot Teachers

The authors would like to acknowledge the following people who field tested preliminary versions of *Exploring Data* and *Predicting with Data* in the schools indicated and whose experiences supported the development of the Teacher's Annotated Editions.

Ellen Bacon
Bedford High School
Bedford, Michigan

Elizabeth Berg
Dominican High School
Detroit, Michigan

Nancy Birkenhauer
North Branch High
School
North Branch, Michigan

Peggy Bosworth
Plymouth Canton High
School
Canton, Michigan

Bruce Buzynski
Ludington High School
Ludington, Michigan

Sandy Clark
Hackett Catholic Central
High School
Kalamazoo, Michigan

Tom Duffey
Marshall High School
Marshall, Michigan

Lonney Evon
Quincy High School
Quincy, Michigan

Carole Fielek
Edsel Ford High School
Dearborn, Michigan

Stanley Fracker
Michigan Center High
School
Michigan Center,
Michigan

Bonnie Frye
Kalamazoo Central High
School
Kalamazoo, Michigan

Raymond Kossakowski
East Catholic High
School
Detroit, Michigan

William Leddy
Lamphere High School
Madison Heights,
Michigan

Dorothy Louden
Gull Lake High School
Richland, Michigan

Michael McClain
Harry S. Truman High
School
Taylor, Michigan

Diane Molitoris
Regina High School
Harper Woods, Michigan

Rose Martin
Battle Creek Central
High School
Battle Creek, Michigan

Carol Nieman
Delton-Kellog High
School
Delton-Kellog, Michigan

Beth Ritsema
Western Michigan
University
Kalamazoo, Michigan

John Schneider
North Branch High
School
North Branch, Michigan

Katherine Smiley
Edsel Ford High School
Dearborn, Michigan

Mark Thompson
Dryden High School
Dryden, Michigan

Paul Townsend
W.K. Kellog Middle
School
Battle Creek, Michigan

William Trombley
Norway High School
Norway, Michigan

Carolyn White
East Catholic High
School
Detroit, Michigan

To the Student

The most often asked question in mathematics classes must be "When am I ever going to use this?" One of the major purposes of *Predicting from Data* is to provide you with a positive answer to this question.

There are several characteristics that this unit has that you may have not experienced before. Some of those characteristics are described below.

Investigations

Predicting from Data consists of five investigations. Each investigation has one, two, or three related activities. After a class discussion introduces an investigation or activity, you will probably be asked to work cooperatively with other students in small groups as you gather data, look for patterns, and make conjectures.

Projects

A project is a long-term activity that may involve gathering and analyzing data. You will complete some projects with a group, some with a partner, and some as homework.

Portfolio Assessment

These suggest when to select and store some of your completed work in your portfolio.

Share and Summarize

These headings suggest that your class discuss the results found by different groups. This discussion can lead to a better understanding of key ideas. If your point of view is different, be prepared to defend it.

Displaying Paired Data

Mary Molar is the Athletic Director at Lindell High School. She keeps records of the growth and abilities of all the athletes. Every fall she collects data on athletes in each of grades 9-12. The data she records for males includes weight, height, bench-press weight, 40-yard dash time, age, and grade-point average (GPA) in school. For females, she records weight, height, leg-press weight, 40-yard dash times, age, and grade-point average. She keeps all this information in a computer. Her computer will print lists of data for groups of students. Ms. Molar wants to organize and display her data so that relationships used to understand how the athletes are progressing will be more obvious.

Activity 1-1 Looking for Relationships

The data collected at Lindell High School for 20 male and 20 female athletes over four years is displayed in Tables 1-8 on pages 5 and 6. Table 1 contains data for the male students when they were in the ninth grade, Table 2 for the same students when they were in the tenth grade, Table 3 for these students in the eleventh grade, and Table 4 for these students in the twelfth grade. Tables 5-8 contain similar data for 20 female athletes collected over four years.

Before you can use the data effectively, you need to become familiar with the data.

● Group Project

Share & Summarize

1. **a.** What is a "bench press"? What is "GPA"? What is a "leg press"?
   See the Teacher's Answer Key.
   b. What is the unit of measure for the data in each category? Be prepared to explain your selections to the class. ***Age – months; Weight – pounds; Height – inches; Bench press – pounds; Leg press – pounds; 40-yd time – seconds; GPA – points***

2. a. Prepare box-and-whisker plots of the weights of male and female athletes in each grade. *See the Teacher's Answer Key.*

b. For each gender, how do the distributions of weights vary from grade to grade? *See the Teacher's Answer Key.*

c. In the ninth grade, what is a typical weight of a male athlete? Did you use the mean or the median? Why? How does the typical weight change from grade to grade? *See the Teacher's Answer Key.*

3. a. Prepare box-and-whisker plots of the heights of male and female athletes in each grade. *See the Teacher's Answer Key.*

b. How do the distributions of heights vary for each gender from grade to grade? *See the Teacher's Answer Key.*

c. In the eleventh grade, what is a typical height of a female athlete? How does this compare with the typical height in the other grades? *See the Teacher's Answer Key.*

Share & Summarize

4. a. Do male athletes vary more in height or in weight as ninth graders? Be prepared to explain your reasoning and choice of statistics to the class. *See the Teacher's Answer Key.*

b. Do female athletes vary more in height or in weight as twelfth graders? Be prepared to explain your reasoning and choice of statistics to the class. *Answers may vary because of the comparison using different units.*

Homework Project

5. a. What is the mean bench-press weight for each of the grades? *129.5 pounds, 143.75 pounds, 158.25 pounds, 173.75 pounds*

b. How does it change as male athletes progress from grade to grade? *It increases by about 14 pounds per year.*

c. Would your response to the previous question change if you used the median? Explain. *No; Bench-press weight still increases, but the increases vary between 10 and 15 pounds per year.*

6. a. What is the median leg-press weight for each of the grades? *150 pounds, 162.5 pounds, 170 pounds, 177.5 pounds*

b. How does it change as female athletes progress from grade to grade? *See the Teacher's Answer Key.*

c. Would your response to the previous question change if you used the mean? Explain. *No; Actual means are 143, 163.25, 172, and 180.25 pounds There is still a dramatic increase between ninth and tenth grade and then steady increases thereafter.*

7. a. Describe the male ninth graders who have the fastest 40-yard dash times. *Sample answer: They tend to be between 68 and 70 inches tall and weigh 160-170 pounds.*

b. Describe the male ninth graders who bench press the most weight. *They are generally the older, heavier, and taller students.*

c. How do the descriptions in part b differ from those in part a? How are they similar? *Answers may vary depending on which characteristics are compared.*

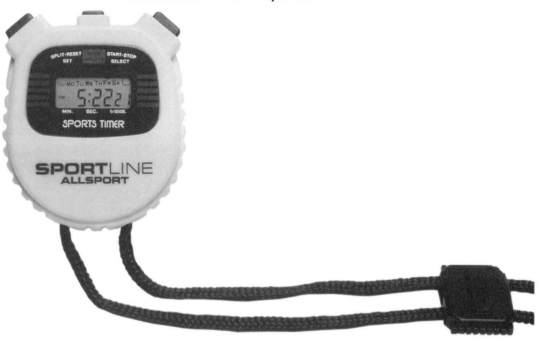

8. Extension Would it be fair to say that younger ninth-grade females run faster or slower than older ninth-grade females? Explain. *See the Teacher's Answer Key.*

Share & Summarize

9. Extension Would it be fair to say that taller ninth-grade females run faster or slower than shorter ninth-grade females? Be prepared to explain your reasoning to the class. *The mean and median times of the six tallest females compared to the six shortest females suggest little, if any, difference.*

10. Extension If Ms. Molar wanted fast-running players for the soccer team, what sort of ninth grader should she look for? *Answers may vary. Primarily she should look for ninth graders with fast 40-yd dash times. Generally both male and female students will be of average height and weight.*

Table 1: Data for Ninth Grade Male Athletes

Student	Age (mo)	Weight (lb)	Height (in.)	Bench Press (lb)	40-yd time (s)	GPA
1	168	125	65	100	7.2	3.2
2	161	130	67	110	7.5	3.7
3	172	150	70	140	6.1	2.8
4	180	170	73	150	6.0	3.0
5	171	168	69	130	5.4	2.3
6	176	200	73	150	7.6	3.1
7	172	155	73	145	6.1	2.6
8	165	143	65	105	5.9	2.8
9	167	156	69	145	5.6	3.8
10	174	169	70	135	5.5	2.1
11	179	170	75	140	6.3	2.7
12	169	125	69	110	5.5	3.0
13	175	135	62	130	6.0	3.6
14	179	178	71	150	5.7	3.6
15	182	190	77	160	6.3	3.1
16	172	157	71	150	5.9	2.5
17	162	120	67	105	5.9	3.1
18	168	140	65	90	5.7	3.2
19	172	160	68	120	5.3	2.8
20	177	155	69	130	5.8	2.4

Table 2: Data for Tenth Grade Male Athletes

Student	Age (mo)	Weight (lb)	Height (in.)	Bench Press (lb)	40-yd time (s)	GPA
1	180	145	67	120	6.9	3.4
2	173	135	68	115	7.3	3.6
3	184	160	73	150	5.9	2.9
4	192	175	74	160	5.7	3.4
5	183	168	70	145	5.2	2.7
6	188	195	73	165	7.1	3.0
7	184	165	75	160	6.2	2.9
8	177	147	66	130	5.8	2.9
9	179	160	70	155	5.4	3.6
10	186	171	71	140	5.3	2.6
11	191	173	77	165	6.0	2.7
12	181	130	70	125	5.4	3.2
13	187	135	62	130	5.8	3.6
14	191	180	72	165	5.6	3.4
15	194	195	77	175	6.1	2.6
16	184	157	72	160	5.1	2.9
17	174	130	69	115	5.5	3.5
18	180	150	66	110	5.7	3.0
19	184	160	69	140	5.2	2.5
20	189	158	70	150	5.8	2.6

Table 3: Data for Eleventh Grade Male Athletes

Student	Age (mo)	Weight (lb)	Height (in.)	Bench Press (lb)	40-yd time (s)	GPA
1	192	165	71	150	6.2	3.6
2	185	145	68	120	7.1	3.8
3	196	171	74	165	5.6	2.5
4	204	180	74	170	5.5	3.6
5	195	170	72	160	5.1	2.9
6	200	196	75	180	6.8	3.3
7	196	170	75	170	6.1	2.5
8	189	155	70	150	5.5	2.8
9	191	165	70	160	5.4	3.6
10	198	176	73	150	5.2	2.8
11	203	179	77	190	5.9	2.9
12	193	140	72	140	5.3	3.4
13	199	145	66	140	5.7	3.6
14	203	185	73	180	5.4	3.2
15	206	200	77	200	5.9	2.8
16	196	160	73	170	5.0	2.8
17	186	140	70	130	5.4	3.5
18	192	155	67	130	5.6	3.1
19	196	165	70	150	5.2	2.7
20	201	169	70	160	5.4	2.9

Table 4: Data for Twelfth Grade Male Athletes

Student	Age (mo)	Weight (lb)	Height (in.)	Bench Press (lb)	40-yd time (s)	GPA
1	204	175	72	160	5.9	3.4
2	197	167	71	160	6.8	3.9
3	208	180	74	170	5.8	2.9
4	216	183	74	200	5.4	3.4
5	207	175	72	180	5.0	3.0
6	212	200	75	210	6.6	3.5
7	208	172	75	175	6.0	2.6
8	201	155	71	160	5.6	2.9
9	203	170	70	170	5.3	3.5
10	210	173	73	170	5.1	2.9
11	215	180	77	200	5.5	3.0
12	205	150	72	145	5.4	3.5
13	211	150	66	130	5.8	3.8
14	215	187	74	190	5.4	3.2
15	218	200	77	220	5.6	2.9
16	208	165	73	190	5.0	2.6
17	198	150	71	150	5.3	3.5
18	204	165	69	160	5.3	3.3
19	208	170	72	170	5.2	2.5
20	213	172	71	165	5.6	2.6

Table 5: Data for Ninth Grade Female Athletes

Student	Age (mo)	Weight (lb)	Height (in.)	Bench Press (lb)	40-yd time (s)	GPA
1	172	127	68	150	6.7	3.2
2	165	104	58	120	6.2	2.7
3	180	130	66	170	6.0	3.6
4	167	107	59	135	6.8	2.5
5	173	111	60	140	5.8	2.3
6	179	125	67	180	5.7	3.7
7	168	109	59	150	5.6	3.5
8	172	110	62	165	5.5	2.6
9	170	132	67	185	6.0	3.1
10	165	90	59	140	5.8	2.4
11	167	97	63	130	5.7	2.7
12	178	115	61	155	5.5	2.9
13	180	130	65	195	5.7	3.3
14	166	102	60	120	6.2	3.7
15	172	117	64	145	6.0	2.3
16	176	111	63	160	5.9	3.5
17	171	135	68	170	5.9	2.8
18	165	95	61	140	5.4	3.1
19	179	121	67	155	5.5	2.1
20	168	103	62	150	5.3	3.4

Table 6: Data for Tenth Grade Female Athletes

Student	Age (mo)	Weight (lb)	Height (in.)	Bench Press (lb)	40-yd time (s)	GPA
1	184	130	68	165	6.5	3.3
2	179	110	59	140	6.1	2.5
3	192	132	66	185	5.8	3.5
4	189	113	59	150	6.3	2.7
5	185	114	61	150	5.7	2.2
6	191	125	67	190	5.4	3.6
7	180	115	60	165	5.6	3.2
8	184	110	64	170	5.3	2.9
9	182	130	67	195	5.8	3.4
10	177	95	60	160	5.6	2.5
11	179	102	64	140	5.5	2.4
12	190	117	61	155	5.2	2.8
13	192	135	65	200	5.6	3.1
14	178	108	62	125	6.1	3.8
15	184	115	65	160	5.9	2.4
16	188	121	64	165	5.9	3.1
17	183	135	68	180	5.7	2.9
18	177	100	63	150	5.2	3.1
19	191	124	67	165	5.4	2.2
20	180	108	63	155	5.1	3.5

Table 7: Data for Eleventh Grade Female Athletes

Student	Age (mo)	Weight (lb)	Height (in.)	Bench Press (lb)	40-yd time (s)	GPA
1	196	135	68	175	6.4	3.0
2	191	112	60	150	6.0	2.3
3	204	139	66	195	5.8	3.6
4	201	115	60	155	6.1	2.9
5	197	112	61	150	5.5	2.4
6	203	125	67	200	5.4	3.5
7	192	120	60	175	5.5	3.4
8	196	124	64	180	5.2	3.0
9	194	132	67	210	5.8	3.4
10	189	98	61	160	5.4	2.7
11	191	104	64	150	5.3	2.3
12	202	115	61	165	5.2	2.9
13	204	135	65	215	5.6	3.2
14	190	110	62	135	5.8	3.8
15	196	115	65	160	5.8	2.5
16	200	117	64	170	5.9	3.2
17	195	138	68	185	5.6	3.0
18	189	102	63	160	5.1	3.0
19	203	125	68	180	5.4	2.3
20	192	110	63	170	5.1	3.4

Table 8: Data for Twelfth Grade Female Athletes

Student	Age (mo)	Weight (lb)	Height (in.)	Bench Press (lb)	40-yd time (s)	GPA
1	208	139	68	190	6.3	2.9
2	203	115	60	160	5.9	2.4
3	216	140	66	205	5.8	3.6
4	213	117	60	160	6.1	3.0
5	209	110	61	165	5.4	2.3
6	215	129	67	210	5.3	3.3
7	204	118	61	180	5.5	3.4
8	208	128	64	190	5.1	3.0
9	206	135	67	225	5.7	3.3
10	201	102	61	170	5.3	2.7
11	203	110	65	150	5.3	2.3
12	214	118	62	170	5.1	3.0
13	216	142	65	220	5.6	3.1
14	202	112	62	140	5.7	3.8
15	208	119	65	170	5.8	2.5
16	212	114	64	175	5.8	3.1
17	207	135	69	200	5.5	3.0
18	201	105	64	160	5.0	2.9
19	215	131	68	185	5.3	2.3
20	204	116	64	180	4.9	3.4

Activity 1-2   Scatter Plots

Materials

 tape measure

 graph paper

 cylindrical objects

 string

Often it is not easy to look at number pairs and "see" whether or not they are associated in some way. Graphing the pairs of data on a coordinate system is one way to organize the data so that associations are easier to "see." Such a graph is called a **scatter plot**.

 **Group Project 1**

1. Below is a procedure to construct a scatter plot for "height" and "armspan" for your class.

Name	Height (in.)	Armspan (in.)

a. Measure, in inches, the height and armspan of each person in your group. Record your data in a table like the one shown at the right.

b. Extend your table and combine your data with that of the other groups.

c. On a piece of graph paper, draw a large ∟ to represent the horizontal and vertical axes, and label the axes. For these data, write *Height (in.)* along the horizontal axis and *Armspan (in.)* along the vertical axis.

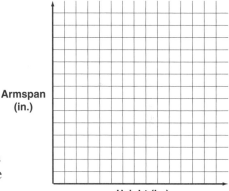

d. Number the tick marks on each axis to make a scale appropriate for the data given.

e. Plot each point (height, armspan) using the pairs of values taken from the class data you collected in part b above as coordinates.

Your graph is a scatter plot of *armspan versus height.*

See students' work. Throughout this unit, scatter plots will be denoted with the first label being the vertical axis and the second label being the horizontal axis.)

2. Does it appear that there is an association between the armspan and height of students in your class? That is, is an increase in one accompanied by an increase in the other, or is an increase in one accompanied by a decrease in the other? If either case exists, we say they are **associated**. Describe any patterns you see in the scatter plot of these data. ***Yes, as height increases, armspan increases also. The armspan is generally within 2 inches of the height.***

3. Are there any clusters of points, that is, a grouping of points close together and somewhat separated from the other points? If there are, describe the characteristics of each cluster. ***Answers may vary.***

4. a. Use this scatter plot to estimate the armspan of a classmate whose height is 65 inches. ***Answers may vary.***

    b. Estimate the height of a classmate whose armspan is 68 inches. ***Answers may vary.***

    c. Explain how you made your estimates using the scatter plot. ***Answers may vary.***

5. Describe how you would make a scatter plot of *height versus armspan*. ***Let height be represented on the y-axis and armspan be represented on the x-axis and regraph the data points.***

    The scatter plot for your armspan versus height data depicts a **positive association** in the data. It is positive because the points of the scatter plot rise as the values on the horizontal axis increase. An increase in one variable is accompanied by an increase in the other.

6. a. Collect from each classmate the distance, in inches, from the ceiling to the tip of the fingers reaching toward the ceiling. ***See students' work.***

Share & Summarize

    b. Make a scatter plot of the distance to the ceiling from fingertip versus height. Label the horizontal axis *Height (in.)* and the vertical axis *Distance to ceiling from fingertip (in.)*. Put a scale on each so that each point can be graphed and so that the points are spread out along the horizontal axis. Describe how you chose the scale for each axis. Be prepared to share your work with the class. ***See students' work. Scales will vary depending upon data. The range of the data should help to influence scale choice.***

Distance to ceiling from fingertip (in.)

Height (in.)

7. Do you think the two variables, height and distance to ceiling from fingertips, are associated? Is the association positive? Give your reasoning. *See the Teacher's Answer Key.*

8. Describe the association as you see it. *See the Teacher's Answer Key.*

9. a. Are there any clusters of points? If so, describe the characteristics of the students in these clusters. *Answers will vary.*

b. Are there any outliers, that is, individual points that are substantially removed from the rest? *Answers may vary.*

10. a. Predict, using the scatter plot, the distance to the ceiling from the fingertip of a student 65 inches tall. *Answers may vary and will depend on the height of the ceiling.*

b. Predict the height of a student whose distance to the ceiling from fingertips is 30 inches. Explain why you chose this value. *Answers may vary.*

Your scatter plot of the distance to ceiling from fingertip versus height data illustrates a **negative association** since as the height of students increases, the distance to the ceiling decreases. In general, the points on the scatter plot get closer to the horizontal axis as the values on the horizontal axis increase.

 Graphing Calculator Activity

You can learn how to use a graphing calculator to make a scatter plot in Activity 1 on page 58.

 Partner Project

11. a. Use computer software or a graphing calculator to make a scatter plot of *Shoe Length (in.) versus Month of Birth*. Recall that the vertical axis is labeled "Shoe Length (in.)" while "Month of Birth" is used to label the horizontal axis. *See students' work.*

b. Does this scatter plot show either a positive or a negative association? Explain your reasoning. *Answers will vary, however, neither a positive nor a negative association should prevail.*

c. How would you label the axes for a scatter plot showing *Month of Birth versus Shoe Length*? *Month of Birth is the y-axis label, and Shoe Length is the x-axis label.*

12. Could you predict the month of birth of a person on the basis of his or her shoe length? Explain. *No; Because of the random scatter of data, no pattern emerges from which to predict shoe length.*

13. Describe the appearance of the points in the scatter plot. *The points are scattered all over the plot.*

The data for shoe length and month of birth show no consistent association. The month of birth of a person may increase or decrease with increasing shoe length. The points are scattered all over the graph. We say that there is **no association**. You cannot make reliable predictions from such data.

14. a. Make a table to record the measurements of the diameter and circumference of the objects you are provided. Wrap a string around each cylindrical object and measure the string to determine the circumference. Then place the object on the tape measure and find the diameter as accurately as possible. Record the measurements in your table. Combine your measurements with those of your classmates. **See students' work.**

b. Use technology to make a scatter plot of *circumference versus diameter*. Which axis should be labeled circumference? **the x-axis**

c. Does the data show any association? Is the association positive, negative, or non-existent? **yes; positive**

d. Describe the scatter plot.
The points appear to lie on a line through the origin.

The *circumference versus diameter* scatter plot showed points that appear to lie in a line. A scatter plot with all the points nearly on a line shows a very strong association. It could be positive or it could be negative, but in either case, it is very strong.

● Group Project 2

15. Sketch three scatter plots each having 15 data points. The first should show a positive association, the second a negative association, and the third little or no association. Compare your plots with that of your group members. Are they similar? Which seems to show the strongest association? **See students' work.**

16. Extension Suppose *A* and *B* represent two variables for which you have collected data. If a scatter plot of the *A versus B* data shows a positive association, what can you conclude about a scatter plot of *B versus A*? Explain your reasoning. **Both scatter plots exhibit a positive association because both are increasing.**

17. Extension Within your group, determine the names of ten rock groups you consider most popular. Each member should make a list of these groups. Each member should then choose a classmate from another group whom they think has similar or dissimilar taste in rock groups. Ask that person to rank your ten groups by placing a 1 after the one liked best, a 2 after the next, and so on. While this is happening, you do the same using your ten groups. Summarize the data in a scatter plot with your ranking as the first coordinate and your classmate's ranking as the second coordinate. Are the two ratings associated? Would you conclude that the two of you have similar or dissimilar taste in rock groups? Explain. **Answers may vary.**

Journal

18. a. Journal Entry What is a scatter plot? **a coordinate graph containing plotted, paired data**

b. Journal Entry Illustrate a scatter plot showing positive association, one showing negative association, and one showing little or no association. **See the Teacher's Answer Key for sample plots.**

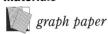
Activity 1-3 Using Scatter Plots to Analyze Data

Computer software and/or graphing calculators simplify the task of constructing scatter plots so you can focus on analyzing the graph, identifying patterns, and communicating your thinking about the data. Recall the Lindell High School data on pages 5 and 6 for males and for females. Choose either males or females for your work in this activity. If a printer is available, print a copy of your scatter plots for future reference and to share with others.

⬤ GROUP PROJECT

1. **a.** Construct a scatter plot for the *bench-press weight* or *leg-press weight* versus *body weight* for Lindell High School ninth graders.
   See the Teacher's Answer Key.

   b. Does it appear that there is an association between the data sets? If so, is it positive or negative? **yes; positive**

1c. You may wish to have students illustrate strong, moderate, and weak associations.

   c. Does the association appear to be strong, moderate, or weak? Explain your answer. *See the Teacher's Answer Key.*

2. Are there any clusters of points ? If so, describe the characteristics of each cluster. *See the Teacher's Answer Key.*

3. Use this scatter plot to estimate the bench-press weight that a 170-pound ninth grade male could lift or the leg-press weight a 115-pound ninth grade female could lift. How did you use the scatter plot to make your estimate? Be specific. *Answers may vary.*

4. Estimate the body weight of a ninth grade male who bench presses 170 pounds or the body weight of a ninth grade female who leg presses 145 pounds using your scatter plot. How did you use the scatter plot to make this estimate? What decisions did you have to make? *See the Teacher's Answer Key.*

Share & Summarize

5. Does your scatter plot show all the points? If not, what points represent more than one item of data? How could you indicate on the scatter plot that a point represented two or more occurrences of an ordered pair in the data? Compare your method with that of a classmate. Which do you prefer? Why? Be prepared to share your methods with the class. *Answers may vary.*

6. Construct a scatter plot for the body weight and 40-yard dash data for eleventh graders included in Table 3 or Table 7 of the Lindell H.S. data. Mark any double or triple points in the manner you chose in Exercise 5. *See the Teacher's Answer Key.*

7. Do you think the two variables, body weight and speed in the 40-yard dash, are associated? Is the association positive or negative? Give your reasoning. ***See the Teacher's Answer Key.***

8. Are there any clusters of points? If so, describe the characteristics of the students represented by these clusters. Are there any outliers? If so, are they outlying on one or on both variables? ***See the Teacher's Answer Key.***

9. Does it make sense to ask for a predicted 40-yard dash time for a tenth grade male weighing 165 pounds or a tenth grade female weighing 115 pounds? If so, do it and explain. If not, explain your reasoning. ***See the Teacher's Answer Key.***

10. a. Refer to the height and GPA data for seniors included in Table 4 or in Table 8. Would you guess that these variables are associated or not? ***Answers will vary. Generally, height and GPA are not associated.***

b. Prepare a scatter plot for GPA versus height for the twelfth grade males or females. ***See the Teachers Answer Key.***

c. Does your scatter plot confirm or disprove your guess? If the plot is associated, describe the association. ***See the Teacher's Answer Key.***

d. Use your scatter plot to predict the GPA of a student who is 68 inches tall. Would you have great confidence in your prediction? Explain. ***See the Teacher's Answer Key.***

11. a. Construct a scatter plot for the twelfth grade age data versus the ninth grade age data. Use Tables 1 and 4 or Tables 5 and 8. ***See the Teacher's Answer Key.***

b. Are these two sets of data positively or negatively associated? Explain your position. ***See the Teacher's Answer Key.***

c. Is the association strong or weak? Explain. ***strong because all the points lie on a line***

d. If you predicted the age of a twelfth grader on the basis of his or her age as a ninth grader, would you have great or little confidence in your prediction? Explain. ***Great confidence because of the extremely strong association between points.***

e. What are the characteristics of this graph? Where do all the points in the scatter plot lie? ***The line possesses a positive slope and there are no outliers. All points lie on a line.***

Homework Project

Share & Summarize

12. Investigate the association of several other sets of data from Tables 1-4 or Tables 5-8. Choose pairs that you think will show association and other pairs that you think would not show association. Were your hunches correct for these data? Be prepared to share your findings with the class. ***Answers will vary.***

 **Journal**

13. Journal Entry Use the data in Tables 1-4 or in Tables 5-8 to respond to the following questions.

    a. Is speed as a twelfth grader associated with speed as a ninth grader? ***yes***

    b. Is speed as a twelfth grader associated with strength as a ninth grader (bench press or leg press)? ***no***

    c. Is strength as a twelfth grader associated with age as a ninth grader? ***yes, but weakly***

    d. Is GPA as a twelfth grader associated with age as a ninth grader? ***no***

    e. Make up a similar question. Construct a scatter plot and then use the graph to respond to the question. ***See students' work.***

14. Extension Suppose all the points graphed in a scatter plot lie on a line. If the association is positive, what could the line look like? If it is negative, what could the line look like? Tell why you chose the lines you did. ***See the Teacher's Answer Key.***

15. Extension If **perfect association** of two sets of data has all the points falling on a line, how could you estimate the strength of an association if all the points did not fall on a line? ***See the Teacher's Answer Key.***

Activity 1-4 How Strong is an Association?

 Materials

software

    The three scatter plots below all show positive association because the points rise from left to right. The scatter plot on the left shows the strongest positive association because it can be enclosed in a narrow oval loop. We say the data represented in the graph has a **correlation** that is positive and large. The next scatter plot can be enclosed in a loop, but it is wider than the first. Thus the correlation is smaller but still is positive. The final scatter plot has a small positive correlation because the loop is quite fat. In general, the correlation is greater when the points of the scatter plot make a long thin loop.

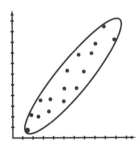 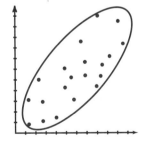 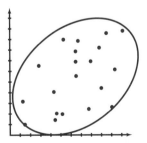

⬤ Partner Project

    1. The three scatter plots at the top of the next page show negative associations. Why are the associations negative? ***The points fall from left to the right.***

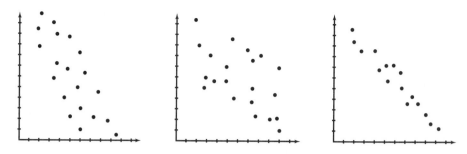

2. Visualize each scatter plot above enclosed with a loop similar to those shown for the variables in the scatter plots on the previous page.

    a. Which scatter plot shows the strongest association? Statisticians say this represents data having a large negative correlation. *the scatter plot on the right*

    b. Which scatter plot shows the weakest association or smallest negative correlation? Explain your reasoning. *See the Teacher's Answer Key.*

3. If the loop you need to enclose a scatter plot is nearly circular, then the association between the variables is low, and the correlation is nearly zero. Which of the scatter plots below appear to have a very small or weak correlation? *See the Teacher's Answer Key.*

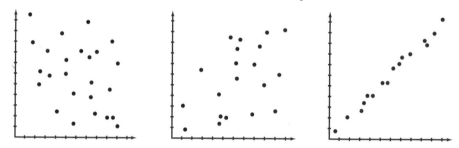

Share & Summarize

4. Correlations are usually reported as numbers between −1 and 1. These are called **correlation coefficients**. A 1 represents perfect positive association, −1 represents perfect negative association, and 0 represents lack of association. Correlation coefficients between 0 and 1 represent positive associations. The stronger the association, the closer the number is to 1. Describe what correlation coefficients between 0 and −1 would mean for the data. Be prepared to share your reasoning with the class. *See the Teacher's Answer Key.*

5. Use software to make a scatter plot for each pair of variables. Print a copy of each scatter plot. *See the Teacher's Answer Key.*

    a. Tenth grade: bench press or leg press versus weight

    b. Twelfth grade: time in 40-yard dash versus weight

    c. Eleventh grade: GPA versus height

    d. Age in grade 12 versus age in grade 10

**Graphing
Calculator
Activity**

You can learn how to
use a graphing
calculator to find the
correlation coefficient
in Activity 2, on page
59.

**Portfolio
Assessment**

A portfolio is
repesentative
samples of your work,
collected over a
period of time. Begin
your portfolio by
selecting an item that
shows something new
you learned in this
investigation.

6. Put loops around each scatter plot in Exercise 5. Estimate what you think the correlation coefficients might be and write them below the appropriate graphs. *See students' work.*

7. Extension Graphing calculators and some computer software that display scatter plots will also calculate the correlation coefficient. Use the technology to find the correlation coefficients in Exercise 5 and compare your estimates in Exercise 6 with these values. *See the Teacher's Answer Key.*

8. Extension Copy and complete the table of movie preferences listed below. Write a 1 for the movie you like the best, 2 for the second best, and so on to 12 for the least liked. Make a scatter plot of the data. Does there seem to be an association between you and your partner's rankings? Explain. *Answers may vary. (You may wish to use movies that are currently showing before assigning.)*

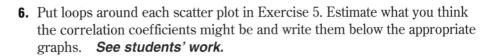

Name of Movie	My Ranking	Partner's Ranking
Home Alone		
Sister Act		
Batman Returns		
Free Willy		
Flintstones		
The Fugitive		
The Beverly Hillbillies		
The Addams Family		
Ace Ventura		
Wayne's World		
A League of Their Own		
Jurassic Park		

9. Extension When ranked data, such as that in the table above, have no ties, the correlation coefficient is the same as that of unranked data. The correlation coefficient is called the **Spearman Rank Order Correlation** and may be found using the following formula.

$$r = 1 - \frac{6(\text{sum of the squares of the differences in the rankings})}{n(n^2 - 1)}$$

In this formula, n represents the number of rankings. Use the data in the table of movie preferences to calculate the Spearman Rank Order Correlation. *See students' work.*

Journal

10. Journal Entry Write a paragraph or two describing what you have learned in this investigation. Include ideas of how you might use this knowledge outside of mathematics class. *See students' work.*

Line Fitting

Ms. Molar, the Athletic Director at Lindell High School, is uncertain about how she can use the data about her athletes. She can now see, using scatter plots, that certain characteristics seem to be associated with others while other pairs seem to show little association. She would like to use data to set goals for her athletes – goals the ninth graders can strive for as they go through school. But when she looks at the scatter plots, she sees a lot of variation in twelfth grade performance given similar ninth grade characteristics. When she talks to the ninth graders, what advice can she give regarding each person's goals for running and lifting weights in the future grades?

What Ms. Molar needs is a way to summarize a scatter plot that lets her predict values for a variable on the basis of the values of another variable. A line can summarize a scatter plot that shows either positive or negative correlation. A **summarizing line** follows the general pattern of the scatter plot and goes through the "middle" of the scatter plot. This means that there are about as many points above the line as below it and that the line contains some of the points. Using a line to summarize a scatter plot is called **fitting a line to the data**.

Activity 2-1 Fitting Lines Visually

Materials

 graph paper

 string

 ruler

The 100-meter dash has been run in the Olympics since 1896. The men's times for each of the years through 1988 are given in the table below.

Men's Olympic 100-Meter Dash Winning Times																						
Year	1896	1900	1904	1908	1912	1920	1924	1928	1932	1936	1948	1952	1956	1960	1964	1968	1972	1976	1980	1984	1988
Time	12.0	11.0	11.0	10.8	10.8	10.8	10.6	10.8	10.3	10.3	10.3	10.4	10.5	10.2	10.0	9.95	10.14	10.06	10.25	9.99	9.92

Source: The World Almanac and Book of Facts 1994

FYI
At the 1992 Olympics, held in Barcelona, Spain, the United States men's team won the gold medal in 8 of the 24 track and field events.

PARTNER PROJECT

1. a. Why are there no times for the early 1940s?
   World War II caused cancellation of the Olympic games.

   b. Why are the times given in hundredths beginning in 1972?
   Due to technological advances, electronic timing began.

c. Study the table. Are the winning times associated with the year of the race? If so, describe how they are associated. ***Yes; Generally, as the year increases, the associated winning time decreases.***

2. a. Construct a scatter plot of winning time versus year of race. (Use computer software or a graphing calculator only if they can print a copy of the scatter plot.) What variable does the horizontal axis represent? What variable does the vertical axis represent? (Make sure you choose scales on the axes so that the points spread out across the first quadrant.) ***See the Teacher's Answer Key.***

b. Do the data appear to be associated in a linear fashion? If so, is the association positive or negative?
Yes; negative.

3. a. Use the scatter plot to estimate the winning 100-meter dash time if the Olympics had been held in 1926.
Sample answer: about 10.6 seconds.

b. Estimate the winning time if the Olympics had been held in 1958.
Sample answer: about 10.3 seconds.

c. If you were to train in the 1930's for the 100-meter race, what time would allow you to be competitive?
about 10.3 seconds.

d. What was the best Olympic 100-meter dash time in the 20th century?
See the Teacher's Answer Key.

e. What was the slowest winning time in the 20th century?
11 seconds in 1904.

4. a. Place a piece of string on the scatter plot you created in Exercise 2 to visually estimate a line passing through the middle of the scatter plot.
See students' work.

b. Now use a ruler to draw this line. This is your **visually-fit line**.
See the Teacher's Answer Key for a sample drawing.

c. Use this line to estimate the probable winning time if a race had been run in 1926 and in 1958. ***Sample answer: 10.7 seconds in 1926 and 10.2 seconds in 1958.***

d. Compare these estimates with those you made in Exercise 3. Are they close? In which case were the estimates more easily made? Why? ***See the Teacher's Answer Key.***

5. The line you drew in Exercise 4 is a **mathematical model** of the data in the table and the scatter plot. It can be used to predict one value of a pair of data when given the other value.

a. Use your line to predict the winning times of the races not run in 1940 and 1944. ***Sample answer: 10.6 seconds and 10.5 seconds.***

b. About what year was a time of 10.5 seconds expected to occur?
around 1944

c. About what year was a time of 10.0 seconds expected to occur?
around 1976

d. Use your line to predict the winning time in 1992. Consult an almanac and find out what the winning time was in 1992. Is your prediction close to the winning time for 1992? How close? ***Sample answer: 9.8 seconds. Actual winning time in 1992 was 9.96 seconds.***

6. **a.** Identify periods of time when the winning times were better than those predicted by your line. ***Sample answer: The points associated with better times will lie above the line.***

   b. Identify periods of time when winning times were worse than those predicted by the line ***Sample answer: The points associated with worse times will lie below the line.***

Share & Summarize

7. **a.** Compare your scatter plot and summarizing line with that of your neighbor. Do they differ? Your response should consider mathematical characteristics such as closeness of the points to the line, number of points on the line, the tilt of the line being similar to the tilt of the scatter plot, and so on. Be prepared to share your findings with the class. ***Answers may vary.***

   b. How different are the values predicted in Exercise 5? ***Answers may vary.***

   c. How different are the periods identified in Exercise 6? ***Answers may vary.***

8. If you extend the line you drew in Exercise 4 in both directions, does it make sense to use it to predict the winning time for a race to be run in the year 2000? in 2020? in 1860? Explain your reasoning. ***Answers may vary. Discuss the appropriateness of predicting winning times 30 years away. Would the time eventually be predicted to be zero or less than zero?***

HOMEWORK PROJECT

9. Examine the data in the table below. Does it appear that the median age of marriage for men and women is related? Why? ***Yes; As the median female age increases, so does the median male age.***

10. **a.** Use computer software or a graphing calculator to construct a scatter plot of male age versus female age. Print a copy of the scatter plot. Do female and male ages at marriage appear to be related? ***See the Teacher's Answer Key for scatter plot. Yes, but there appears to be one outlier.***

Median Age at First Marriage

Year	Females	Males
1900	21.9	25.9
1910	21.6	25.1
1920	21.2	24.6
1930	21.3	24.3
1940	21.5	24.3
1950	20.3	22.8
1960	20.3	22.8
1970	20.8	23.2
1980	22.0	24.7
1990	23.9	26.1

Source: Department of Commerce. Bureau of Census

b. Visually estimate a summarizing line and draw it with a ruler.
See the Teacher's Answer Key for sample answer.

11. a. Use the line to estimate the age of a man marrying a woman who is 21 years old; 22.5 years old.
Sample answer: About 24; 26 years.

b. Use the line to estimate the age of a woman marrying a man who is 23 years old; 24 years old.
Sample answer: About 20; 21 years.

12. Make a scatter plot of median male's age of first marriage versus year. Can you visualize a line summarizing this scatter plot? Draw your summarizing line or explain why it would be inappropriate to do so. ***See the Teacher's Answer Key.***

Share & Summarize

13. If a scatter plot does not show an association between the variables, then no line can be used to summarize the data, and none should be drawn. For each of the six scatter plots below and on the next page, decide whether a line should be used to summarize the data. Trace the axes of each scatter plot on a separate piece of paper and then draw the line you think best fits the data. Then predict several values of each variable by reading corresponding values of the other variable from the line. Be prepared to share your work with the class. ***See students' work.***

a. male ninth grade weight versus height

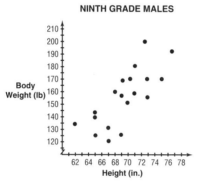

b. female eleventh grade weight versus height

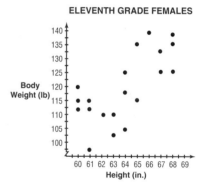

c. male eleventh grade bench press versus time in 40-yard dash

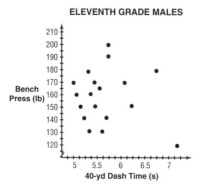

d. female leg press for twelfth graders versus leg press for ninth graders

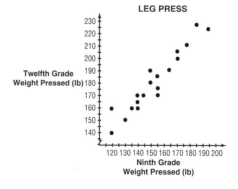

e. female time in 40-yard dash for twelfth graders versus time in 40-yard dash for ninth graders

f. male tenth grade age versus tenth grade GPA

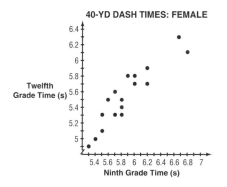

40-YD DASH TIMES: FEMALE

Twelfth Grade Time (s) vs Ninth Grade Time (s)

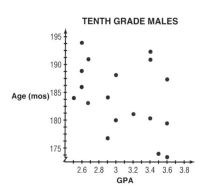

TENTH GRADE MALES

Age (mos) vs GPA

Journal

From this point on, answers for questions asking for summarizing lines or requiring median-fit lines will use the Data Insights program. This program is available from Wings for Learning/Sunburst.

14. Journal Entry

   a. Do you think a line is a good mathematical model for the winning Olympic 100-meter dash times? Explain your position and reasoning. *See the Teacher's Answer Key.*

   b. If you could summarize these data with a different kind of graph, what would it look like when drawn in the scatter plot? Explain its characteristics and why it has those characteristics. *See the Teacher's Answer Key.*

15. Extension

   a. Using computer software, generate other scatter plots for Ms. Molar's data that would be useful to her in advising ninth-grade athletes, both male and female, regarding goals for speed in the 40-yard dash and weight lifted in the bench press or leg press by the time they are in twelfth grade. *See students' work.*

   b. Summarize each set of data with a line. *See the Teacher's Answer Key.*

   c. Use the lines to advise a ninth grader about goals. *See students' work.*

Journal

16. Journal Entry

   a. Describe how you visually fit a line to a scatter plot. *See the Teacher's Answer Key.*

   b. Why are lines fitted to scatter plots? *to summarize data; to predict missing or unknown information*

   c. Should every scatter plot be fitted with a line? Explain your position. *No. If a scatter plot does not show an association between the variables, then no line can be used to summarize the data, and none should be drawn.*

Activity 2-2   How Good is Your Fit?

Materials

 graph paper

 ruler

Women have participated in the Olympic high jump since 1928. Below is a table of the heights jumped by women in the high jump event through 1988.

Women's Olympic High Jump Winning Heights

Year	Country	Height
1928	Canada	5 ft 3.000 in.
1932	United States	5 ft 4.250 in.
1936	Hungary	5 ft 3.000 in.
1948	United States	5 ft 6.125 in.
1952	South Africa	5 ft 5.750 in.
1956	United States	5 ft 9.250 in.
1960	Romania	6 ft 0.750 in.
1964	Romania	6 ft 2.750 in.
1968	Czechoslovakia	5 ft 11.750 in.
1972	West Germany	6 ft 3.625 in.
1976	East Germany	6 ft 4.000 in.
1980	Italy	6 ft 5.500 in.
1984	West Germany	6 ft 7.500 in.
1988	United States	6 ft 8.000 in.

Source: Information Please Almanac, 1994

● GROUP PROJECT

1. a. What was the least winning height? When was it a winner?
5 foot 3 inches; 1928 and 1936

b. What was the greatest winning height? When was it a winner?
6 feet 8 inches; 1988

c. What is the mean winning height?
5 feet 11.375 inches

d. What is the median winning height?
6 feet 0.25 inches

2. a. Construct a scatter plot of winning height versus year. Which axis is labeled "Year", and which axis is labeled "Winning Height"? ***See the Teacher's Answer Key.***

b. Is the association positive or negative? ***positive***

c. Is the scatter plot linear in nature? ***yes***

d. Visually determine a summarizing line and draw it with a ruler. ***See students' work.***

e. How many points are above the line? How many are below the line? How many are on the line? ***Answers may vary. There should be close to an equal number above and below the line.***

3a. Check to see that students measure vertical distance, not perpendicular distance.

3. a. Use a centimeter ruler to measure the vertical distance from each point in the scatter plot to the line you drew in Exercise 2d. This distance is called the **residual** of the linear model of the data. Record each residual. ***Answers may vary.***

b. What is the greatest residual? ***Sample answer: 1.2 cm.***

c. What is the least residual? ***Sample answer: 0 cm.***

d. In terms of inches, what heights do the residuals in parts b and c represent? ***See the Teacher's Answer Key.***

Share & Summarize

4. Add the residuals determined in Exercise 3 and find their mean. This **mean residual** is a measure of how well your line fits the data. If the mean is small, then the line is a good fit to the data. A larger mean suggests the line is not a good fit to the data.. How does your mean compare with that of a neighbor? Be prepared to share your reasoning with the class. ***Answers may vary.***

5. a. Try to summarize the data in the scatter plot in Exercise 2 with another line that you think will improve the fit. ***See students' work.***

Share & Summarize

b. Calculate the mean residual for this line and compare it with your original value. Did you improve on your original fit? Be prepared to share your strategies and findings with the class. ***Answers may vary.***

6. a. Using the two lines of Exercises 2 and 5, predict the winning women's high jump for 1992 and for 1996. ***Answers may vary depending on students' lines. 82.5 inches***

b. Do the predictions differ? If so, by how much? ***Answers may vary depending on the lines chosen.***

c. Which prediction do you have the most confidence in? Explain. ***See the Teacher's Answer Key.***

d. Compare your prediction for 1992 with the actual winning height. How close was your prediction? ***Answers will vary. Actual winning height in 1992 was 6 feet 7.5 inches.***

HOMEWORK PROJECT

Journal

7. Journal Entry Why do you think it is important to find the summarizing line that fits the data? *It is important to get the line that best fits the data in order to produce reliable predictions based on that line.*

FYI
In the 1992 Olympics, the United States women's swim team won a gold medal in 5 of the 15 races.

8. Women began swimming the 100-meter butterfly in the Olympic Games in 1956. The winning times for 1956 and the next eight competitions are given in the table below.

Women's 100-meter Butterfly

Year	Country	Time (s)
1956	United States	71.00
1960	United States	69.50
1964	United States	64.70
1968	Australia	65.50
1972	Japan	63.34
1976	East Germany	60.13
1980	East Germany	60.42
1984	United States	59.26
1988	East Germany	59.00

Source: Information Please Almanac, 1994

Portfolio Assessment

Select one of the assignments from this investigation that you found especially challenging and place it in your portfolio.

a. Use these data to predict the winning time in the 1992 Olympic Games. *Sample answer: 58.75 seconds.*

8b. Discuss the accuracy of predictions when there are few data points for references.

b. Are you confident of your prediction? Explain. *Answers may vary.*

c. Compare your predicted winning time for 1992 with the actual winning time. *Answers may vary. Actual winning time in 1992 was 58.62 seconds.*

d. Predict the winning time for 1996. Are you confident of your prediction? *See the Teacher's Answer Key.*

9. a. Describe one way used to determine how well a line fits data in a scatter plot. *Determine the mean residuals with respect to the line. Relatively small means imply a better fit than larger means.*

b. What is a residual? *The vertical distance from a point to a summarizing line in a scatter plot.*

Journal

10. Journal Entry Write a paragraph or two describing what you have learned in this investigation. Include ideas of types of data that have summarizing lines that are a good fit to the data.
See students' work.

Median-Fit Line

Summarizing the data shown in a scatter plot with a line is not an exact process. As you have seen, several lines may be used to summarize the same data. Some of these lines summarize better than others in the sense that the mean of the residuals is smaller.

Statisticians typically use two methods to draw lines that summarize data. In this investigation, you will use the **median-fit line**. It is easily drawn by hand for small sets of paired data. For larger sets of data, a computer should be used.

Materials

 ruler

 tracing paper

 string

 calculator

 software

Activity 3-1 Finding and Using Median-Fit Lines

GROUP PROJECT

1. The table below lists the death rates (deaths per 1000) for United States citizens from 1910 through 1990. What pattern do you see in the data? *Sample answer: The rate is decreasing but at a slower rate the last 40 years.*

United States Deaths per 1,000 Citizens																
Year	1910	1915	1920	1925	1930	1935	1940	1945	1950	1955	1960	1965	1970	1975	1980	1985	1990
Death Rate	14.7	13.2	13.0	11.7	11.3	10.9	10.8	10.6	9.6	9.3	9.5	9.4	9.5	8.8	8.7	8.7	8.6

Source: Information Please Almanac, 1994

2. A computer-generated scatter plot is shown below. Does the data seem to be linear? Represent a visually-fit summarizing line by placing a piece of string on the plot. Be prepared to share your reasoning with the class. *See the Teacher's Answer Key.*

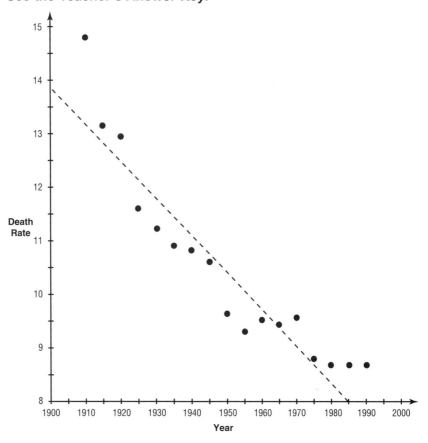

3. The procedure for drawing a median-fit line is described below. Trace the scatter plot above and carry out each of the following instructions on your traced copy.

a. Count the number of data points and divide that total by 3. You will use this number to divide the data points into three sets. If the total is divisible by 3, then the three sets have equal numbers of points. If, however, the total is not divisible by 3, choose the three numbers such that two are equal and larger or smaller than the third. In this example, you have 17 points, which is not divisible by 3, so you choose "thirds" of 6, 6, and 5.

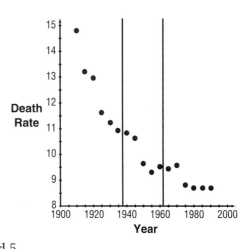

b. Draw two vertical lines on the scatter plot so that the points are divided into three equal sections, or so that the leftmost and rightmost sections are equal and greater than the middle section. For these data you should have sections of 6, 5, and 6, from left to right.

c. The next step is to find the **median point** for each section. The *x*-coordinate of this point is the median of the *x*-coordinates of all the points in the section. The *y*-coordinate is the median of the *y*-coordinates of the points.

Since there are six points in the left section, there are six *x*-values and six *y*-values to use in finding the median point. The median of each set is halfway between the third and fourth values. Find these values. **(1922.5, 12.35)** The median point can be found geometrically by placing a ruler horizontally at the bottom of the section and moving it up until it is halfway between the third and fourth points. Draw a short segment here.

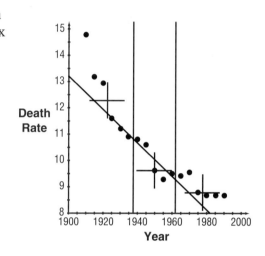

Place the ruler vertically at the leftmost edge of the section. Move it to the right until it is halfway between the third and fourth points. Mark a short segment.

The median point of this section is the point where the two short segments intersect.

d. Repeat step c for the other two sections of the scatter plot. In this way you have identified three points that are median points for the three sections of the scatter plot.

e. Place the ruler on the median points in the left and right sections. This determines the slope of the summarizing line. Now slide the ruler one-third the distance to the point in the middle section, keeping the ruler parallel to it original position. Draw this line. The result is the *median-fit line* for these data. ***See above.***

4. Compare your median-fit line with that of your neighbor. Are they nearly the same? If not, how are they different and why are they different? ***The lines should be very similar. Any variation would result because of differences in plotting the three median points or estimating the $\frac{1}{3}$ distance toward the center median point.***

5. a. On the basis of your median-fit line, predict the death rate in 1900. *Sample answer: approximately 13.6 deaths per thousand citizens*

b. What is the predicted death rate in 1950? How far off is the prediction? *See the Teacher's Answer Key.*

6. On the basis of your median-fit line, what is a prediction for the death rate for 1978? for 1989? for 1995? for 2000? *Sample answers: Approximately 8.4, 7.8, 7.3, 6.9 per 1000 citizens*

7. a. Is the median-fit line a good model of the death rate data? Explain your reasoning. *See the Teacher's Answer Key.*

b. Are there portions of the scatter plot that it fits well? not so well? What are they? *It appears to fit the 1920-1970 data fairly well, other data not so well.*

c. What would be the effect of constructing a median-fit line using only 1930-1990 data? *Sample answer: The slope of the line would be less steep.*

8. a. Compare your visually-fit line to the median-fit line. Are they parallel? *Answers may vary.*

b. Compute the mean of the residuals for each summarizing line. Which is the better fit? *Answers may vary.*

 Share & Summarize

c. Which summarizing line would you have more confidence in if you were to make predictions for the future? Why? Be prepared to explain your reasoning to the class. *Sample answer: The line with the smaller residual mean because it fits the data better.*

 **PARTNER PROJECT**

Graphing Calculator Activity

You can learn how to use a graphing calculator to find median-fit lines in Activity 3 on page 60.

9. Use technology to find the median-fit line for the men's 100-meter dash winning times in the Olympic Games from 1896 through 1988 found on page 16. Compare this with your visually-fit line in Exercise 4 on page 17. *See the Teacher's Answer Key. Answers may vary.*

10. a. Use technology to find the median-fit line for height versus weight of the ninth-grade males or ninth-grade females found on pages 5 and 6. *See the Teacher's Answer Key.*

b. If a male weighed 165 pounds or a female weighed 118 pounds, predict their heights. *Sample answer: About 72 inches for the male and about 64 inches for the female.*

c. If a male or female were 66 inches tall, how much would (s)he weigh? *See the Teacher's Answer Key.*

d. Find the mean of the weights and the mean of the heights for females or males. Plot the point (mean of weights, mean of heights). How close is this point to the median-fit line? *Females: (113.5, 63); Males: (154.8, 69.4) These points are extremely close if not on the median-fit line.*

11. **a.** Use technology to find the median-fit line for height versus weight of eleventh-grade males or eleventh-grade females found on pages 5 and 6. ***See the Teacher's Answer Key.***

   b. What is the predicted height of a 165-pound male or a 118-pound female?   ***Sample answer: About 71 inches for the male, and about 63 inches for the female.***

   c. What is the predicted weight of a 6-foot 2-inch male or a 5-foot 4-inch female?   ***Sample answer: About 185 pounds for the male and about 122 pounds for the female.***

12. **a.** Use technology to construct the median-fit line for twelfth grade versus ninth grade bench-press or leg-press weight lifting using the data found on pages 5 and 6. Identify all multiple points on your scatter plot. ***See the Teacher's Answer Key.***

   b. What bench-press or leg-press weight goal would you suggest to a ninth grade athlete who now presses 130 pounds? 160 pounds? 185 pounds? ***See the Teacher's Answer Key.***

   c. Find the mean of the ninth and twelfth grade bench-press or leg-press weights. Plot the point (mean ninth grade weight, mean twelfth grade weight). How close is this point to the median-fit line?   ***See the Teacher's Answer Key.***

13. **a.** Use technology to construct the median-fit line for the times in the 40-yard dash for ninth grade versus twelfth grade males or females found on pages 5 and 6. Identify all multiple points on the scatter plot. ***See the Teacher's Answer Key.***

Share & Summarize

   b. What time goal, as a senior, would you suggest for a ninth grade athlete who could run the 40-yard dash in 6.6 seconds? in 5.8 seconds? in 6 seconds? Be prepared to share your findings with the class. ***See the Teacher's Answer Key.***

14. **a.** Use technology to construct the median-fit line for the Women's Olympic High Jump Winning Heights on page 21.   ***See the Teacher's Answer Key.***

14b. The actual winning height in 1992 was 79.5 inches.

   b. Predict the winning height for the 1992 and 1996 Olympic Games. ***1992: 81.25 inches; 1996: 82.5 inches***

   c. If you used this line to predict for the year 2020, what height would you get? Is this a reasonable prediction? Explain.   ***Sample answer: 89.5 inches. Answers will vary.***

FYI
FYI
Vessels for deep-sea exploration must have complete life-support systems aboard and be able to withstand the enormous pressures of such depths, which range up to 100,000 pounds per square inch.

15. Extension The areas and depths of the largest oceans and seas are given in the table below.

Name	Area (1000's of sq. miles)	Average depth (ft)	Greatest depth (ft)
Pacific Ocean	64,000	13,215	36,198
Atlantic Ocean	31,815	12,880	30,246
Indian Ocean	25,300	13,002	24,460
Arctic Ocean	5,440	3,953	18,456
Mediterranean	1,145	4,688	15,197
Caribbean Sea	1,050	8,685	22,788
S. China Sea	895	5,419	16,456
Bering Sea	885	5,075	15,659
Gulf of Mexico	615	4,874	12,425
Okhotsk Sea	614	2,749	12,001
E. China Sea	482	617	9,126
Hudson Bay	476	420	600
Japan Sea	389	4,429	12,276
Andaman Sea	308	2,854	12,392
North Sea	222	308	2,165
Red Sea	169	1,611	7,254
Baltic Sea	163	180	1,380

Source: Information Please Almanac, 1994

a. Locate each of these oceans and seas on a globe or map.

b. Where are the deepest areas located?
The deepest areas are located in the Pacific and Atlantic Oceans.

c. Use a statistical software package to construct scatter plots for average depth versus area, greatest depth versus area, and greatest depth versus average depth. **See students' work.**

d. Visually examine each of these scatter plots. Are they difficult to interpret? Are there clusters of points that are "outliers" from the rest? If so, identify them in each scatter plot. ***See the Teacher's Answer Key.***

e. Eliminate any outlier points from each scatter plot and reconstruct. Are relationships more easily observed? Construct the median-fit lines for each scatter plot, if it is appropriate to do so. ***See the Teacher's Answer Key.***

f. Determine an equation describing each line. You may use whatever means you need to find the equation. Use each equation to predict the average depth and the maximum depth of a hypothetical sea with an area of 350,000 square miles, 700,000 square miles and 2,000,000 square miles. ***See the Teacher's Answer Key.***

16. Extension

a. How could you use means to construct a summarizing line to use in place of the median-fit lines? Describe the procedure to use and illustrate it for the United States Birth Rate data given in the table. The rates are per 1000 people living in the United States.
See the Teacher's Answer Key.

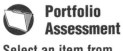

Portfolio Assessment

Select an item from this investigation that you feel shows your best work and place it in your portfolio. Explain why you selected it.

Birth Rates in the U.S.

Year	Rate
1910	30.1
1915	29.5
1920	27.7
1925	25.1
1930	21.3
1935	18.7
1940	19.4
1945	20.4
1950	24.1
1955	25.0
1960	23.7
1965	19.4
1970	18.4
1975	14.8
1980	15.9
1985	15.8

Source: Information Please Almanac, 1991

b. Historically, what explanation is there for the rates in the 1930's and the rate in the 1950's? According to your linear model, what was the expected rate in those periods? ***The early 1930s was a time of great economic depression, which would help to explain the low birth rate. The 1950s was a decade following World War II, which would explain the increased birth rate.***

Linear Patterns

I n the last three investigations, you explored methods that would be helpful to Ms. Molar in making sense of student data at Lindell High School. These methods included visual displays and summarizing lines. A summarizing line made it easy to make a reasonable prediction of the future, but we could not be sure the prediction was exactly right. Was your prediction for the 1992 100-meter dash time exactly the same as the winning time? Some situations with which you are familiar give rise to patterns for which a line can be fitted exactly. As a result, predictions can be exact also.

Activity 4-1 Representing Linear Patterns

Materials

graph paper

ruler

calculator

● PARTNER PROJECT

Mega-Hit Video charges $3.00 to rent a video tape for a day and $9.00 to rent a VCR for a day. Juanita and Cedric are planning weekend video parties. Juanita will feature horror movies, and Cedric will show all action tapes.

1. Copy and complete the table of rental costs of the tapes for Juanita shown below.

Number of Tapes (x)	0	1	2	3	4	5	6	7	8
Cost of tapes (in Dollars) (y)	0	3	*6*	*9*	*12*	*15*	*18*	*21*	*24*

2. a. On a sheet of graph paper, make a scatter plot for cost of tapes versus number of tapes.   ***See the Teacher's Answer Key.***

b. Do the points appear to have a pattern? If so, describe it.   ***Yes. They appear to lie on a line.***

c. Draw the median-fit line for these points.   ***See students' work.***

d. How good is the fit of the median-fit line?   ***It fits all of the points perfectly. That is, all the points are on the line.***

3. a. What are the coordinates of the point where the line and the *x*-axis intersect?   ***(0,0)***

b. What are the coordinates of the point where the line and the *y*-axis intersect?   ***(0,0)***

c. Can the line intersect either axis in more than one point? Explain.   ***No. Since the median-fit line and an axis are two distinct lines and two distinct lines intersect at one point, it is impossible to have more than one point of intersection.***

4. a. What would be the rental cost if Juanita rented 10 tapes for a day?
$30
   b. What would be the rental cost if Juanita rented 50 tapes?   *$150*

   c. Would it make sense for her to rent 50 tapes? Explain.   *See the Teacher's Answer Key.*
   d. What phrase describes the rental cost if Juanita rented *x* tapes?
$3 times the number of tapes rented, x.

5. a. Write, in words, a rule that seems to describe the relation between the number of tapes rented, *x*, and the total cost of the rental, *y*.   *See the Teacher's Answer Key.*
   b. Compare your description with that of a neighbor. Are they identical? similar? Do they say the same thing?   *Answers may vary. They should "mean" the same thing.*
   c. Work together to produce a rule in words you both think is accurate.
See students' work.
   d. Translate your rule into an equation, where *y* represents the cost and *x* represents the number of tapes rented.   *y = 3x*

6. a. Use your rule to find the cost of renting 15 tapes.   *$45*

   b. How much would it cost you to rent 26 tapes?   *$78*

Share & Summarize

   c. How many tapes could you rent if you had $25 to spend? Be prepared to share your findings with the class.   *8 tapes*

● GROUP PROJECT

7. a. Cedric does not have a VCR, so he must rent one. How much rental will Cedric pay to rent a VCR alone for one day? one tape and a VCR?
$9 for the VCR alone; $12 for the VCR and one tape
   b. Copy and complete the table for Cedric's cost to rent movies and one VCR for one day.

VCR and Number of Tapes (*x*)	0	1	2	3	4	5	6	7	8
Cost of VCR and Tapes (in Dollars) (*y*)	9	12	*15*	*18*	*21*	*24*	*27*	*30*	*33*

8. a. On a sheet of graph paper, make a scatter plot for cost (*y*) versus VCR and number of tapes (*x*).   ***See the Teacher's Answer Key.***

b. Describe any pattern(s) you see in the points.   ***The points appear to lie on a line.***

c. Draw the median-fit line for these points.   ***See students' work.***

d. How good is the fit of the median-fit line to the points?   ***It fits all of the points perfectly.***

e. What is the correlation of these data points?   ***The correlation is 1.***

9. a. What are the coordinates of the point where the median-fit line and the *y*-axis intersect?   ***(0,9)***

b. What does the value of the first coordinate represent? What does the value of the second coordinate represent?   ***See the Teacher's Answer Key.***

10. a. Does the median-fit line intersect the horizontal or *x*-axis?   ***not in the first quadrant***

b. If you extend the axis and the median-fit line, do they intersect?   ***yes***

c. What are the coordinates of the point in which the median-fit line intersects the horizontal axis?   ***(−3,0)***

Share & Summarize

d. Describe how you determined the coordinates in part c. Your description should be written so that a classmate could use it to determine the coordinates. Be prepared to share your description with the class.   ***Answers may vary.***

11. a. What would be the rental cost if Cedric rented a VCR and 12 tapes for a day?   ***$45***

b. What would be the rental cost if Cedric rented a VCR and 35 tapes for a day?   ***$114***

c. Would it make sense for him to rent 35 tapes? Explain.   ***See the Teacher's Answer Key.***

d. What would be the rental cost if Cedric rented a VCR and *x* tapes?   ***9 + 3x***

12. a. Write, in words, a rule that describes the relation between the number of movies rented, *x*, and the total cost of the rental, *y*, including one VCR.   ***The total cost of the rental, y, equals $9 for the VCR and $3 times the number of tapes rented, x.***

b. Compare your description with that of a classmate. Are they identical? similar? Do they say the same thing?   ***Answers may vary. (Sentences should represent the same thing.)***

c. Work together to produce a written description you both think is accurate and clear.   ***See students' work.***

d. Express your written description as an equation.   ***y = 9 + 3x***

13. a. Use your rule to find Cedric's cost for renting one VCR and 15 tapes. *$54*

b. How much would it cost to rent the VCR and 26 tapes? *$87*

c. In addition to the VCR, how many tapes could Cedric rent if he had $25 to spend? *5 tapes*

14. The speed limit on many U.S. highways is 65 miles per hour. How many miles would you travel if you drove at this speed limit for 1 hour? for 2 hours? *65 miles; 130 miles*

15. Copy and complete the table for miles traveled in hours at 65 miles per hour shown below.

Hours Driven (x)	1	2	3	4	5	6	7	8	9
Miles Traveled (y)	65	130	195	260	325	390	455	520	585

16. a. Make a scatter plot for miles traveled versus hours driven. You may wish to use a software package to complete the scatter plot. *See the Teacher's Answer Key.*

b. Draw the median-fit line for the scatter plot. *See the Teacher's Answer Key.*

17b. *Discuss whether it is possible and/or reasonable to drive 15 hours at a constant rate of 65 miles per hour.*

17. a. Use your graph to predict the miles traveled at 65 miles per hour if you drive 15 hours. *975 miles*

b. Are you confident in your prediction? Why? *Answers may vary.*

18. *Depending upon the accuracy of the graph, student's answers may be approximate.*

18. Use your graph to find the miles traveled for the given hours of driving.

a. 10 hours   *650 miles*

b. 2.5 hours   *162.5 miles*

c. 1.75 hours   *113.75 miles*

d. 3.25 hours   *211.25 miles*

19. a. Write an equation for the relationship between the number of hours driven, x and number of miles traveled, y.   *y = 65x*

b. Use your equation to find the miles traveled in each part of Exercise 18. Be prepared to share your findings with the class. *650 miles; 162.5 miles; 113.75 miles; 211.25 miles*

Share & Summarize

● HOMEWORK PROJECT

20. Extension

    a. How could you use your graph to find the hours required to travel 300 miles at 65 miles per hour? What do you find? ***See the Teacher's Answer Key.***

    b. Use your equation to find the hours required to drive 400 miles at 65 mph. Describe the procedure you used. ***Approximately 6.15 hours required. Answers will vary.***

21. Extension

    a. Cyndi, a salesperson at Betty's Pretty Good Boutique, works on commission. She makes $300 a week plus a commission of 5% of her total sales. How much does Cyndi earn in a week when she sells $4000 worth of clothing? ***$500***

    b. Make a table showing Cyndi's weekly earnings for each additional $1000 in sales from $0 sales through $10,000 in sales. ***See the Teacher's Answer Key.***

    c. Use your table to make a scatter plot for weekly earnings versus total sales. ***See the Teacher's Answer Key.***

    d. Draw the median-fit line. Where does it intersect the axes? ***See the Teacher's Answer Key.***

22. Extension

    a. Write an equation for the relationship between Cyndi's weekly earnings if y is her earnings and x is her total weekly sales. ***y = 300 + 0.05x***

    b. What are Cyndi's least weekly earnings? When does it occur? ***$300; When she sells nothing.***

    c. What is Cyndi's weekly earnings if her sales are $2350? ***$417.50***

    d. What were Cyndi's sales if she had earnings of $532.50? ***$326.63***

23. Extension

    a. If Cyndi's salary were raised to $350 a week plus 5% commission on weekly sales, construct a graph of her weekly earnings for weekly sales from $0 to $10,000. ***See the Teacher's Answer Key.***

    b. How is this graph similar to the graph in Exercise 21? How is it different? ***See the Teacher's Answer Key.***

    c. Write an equation describing Cyndi's new weekly earnings. How is it similar to the one you wrote in Exercise 22 a.? How is it different? ***y = 350 + 0.05x; The last term is the same; The number after the equal sign, which represents the base salary, changed.***

Activity 4-2   Linear Equations

graph paper

ruler

calculator

software

● PARTNER PROJECT 1

1. a. John Clark sells appliances at Lowland Appliance Store. He earns $250 a week and a commission of 10% on all that he sells. Suppose y is his weekly earnings including the 10% commission on x dollars of sales. Find his earnings if x is $1500.   ***$400***

b. One week John's manager, Sheila Deppler, told John that he had sold $2500 in goods, but that Mr. Bumstead had returned $3000 in goods previously sold. What is John's sales total for the week?   **−*$500***

c. What are John's earnings for the week described in Exercise 1b? ***$200***

2. Copy and complete the table below of John's weekly sales and his earnings.

Sales: x	0	4000	5000	300	-500	2100	125	-4000	1300
Earnings: y	250	650	***750***	***280***	***200***	***460***	***262.50***	***−150***	***380***

3. a. Make a scatter plot for earnings versus sales on a sheet of graph paper. Choose your scale carefully. Label the axes and draw the median-fit line. ***See the Teacher's Answer Key.***

b. Describe the scale you chose for the grid. How did you label points to the left of 0 on the horizontal axis? How did you label points below 0 on the vertical axis?   ***See the Teacher's Answer Key.***

● Share & Summarize

c. Compare your answer to part a and part b with that of your neighbor. Between the two of you, decide on a way to plot points where one or both of the coordinates are negative numbers. Be prepared to share your method with the class.   ***Answers may vary.***

● GROUP PROJECT

4. a. Write a rule in the form of an equation that gives John's earnings, y, in terms of his base salary and his commission of 10% on sales of x dollars. ***$y = 250 + 0.10x$***

b. Use the equation to determine John's earnings for sales of $240, $850, $2340, −$440, and −$672.   ***$274; $335; $484; $206; $182.80***

c. Do the points calculated in part b lie on the median-fit line of Exercise 3a?   ***Yes. Students will estimate locations. Computer software may be effectively used here to show all points.***

5. a. The equation you wrote in Exercise 4a is an example of a linear equation. Describe how this equation is used to calculate John's earnings. ***See the Teacher's Answer Key.***

b. Why do you suppose the equation is called a "linear" equation? Be prepared to share your reasoning with the class. ***because the points that satisfy the equation lie on a line***

6. a. The cost of operating a cellular telephone each month is given by a linear equation. If the base cost is $25 per month and $30 for each hour of use, write the linear equation giving the cost in terms of the number of hours of operation. $c = 25 + 30h$

FYI
From 1984 to 1994, the number of cellular phone subscribers has increased from 125,000 to over 16 million, while the average cost has gone from $275 a month to $72 a month.

b. Make a table that includes seven pairs of values for hours and cost per month for the use of a cellular telephone. ***See the Teacher's Answer Key.***

c. Graph your seven pairs of values on graph paper. ***See students' work.***

d. Do the points lie on a line? If so, draw it. ***Yes. See students' work.***

7. The perimeter *(P)* of a rectangle is the distance around the rectangle. For a rectangle with one dimension that is 10 inches long, the table gives the perimeter for several values of the other side, *s*.

Side: *s*	4	15	6	9	21	2	18	10
Perimeter: *P*	28	50	32	38	62	24	56	40

a. Draw a rectangle and label the sides when *s* = 4 and *P* = 28.

b. Find an equation relating the perimeter, *P*, to the length of the side, *s*. $P = 20 + 2s$

c. Graph the points on graph paper. Do they lie on a line? If so, draw it. ***Yes. See the Teacher's Answer Key.***

d. At what point does the graph intersect the vertical axis? ***(0, 20)***

e. In your equation in part b, if *P* is 18, what is the value of *s*? Does this value for *s* make sense in this situation? Why? ***See the Teacher's Answer Key.***

8. a. The base, *b*, of a triangle is fixed at 10 units. The height of the triangle is represented by *h*. Let *A* represent the area of the triangle. Copy and complete the table of values relating the height and area. Remember that the area of a triangle is given by $A = \frac{1}{2}bh$.

Height: *h*	8	10	4	7	6	3	5	1.2	x
Area: *A*	40	50	20	35	**30**	**15**	**25**	**6**	**5x**

b. Write an equation relating the area, *A*, and the height, *h*. Check your equation for several pairs in the table. ***Answers may vary. Most students will write A = 5h.***

c. Use graph paper to graph the pairs in the table. Is the relation linear? If so, draw the line. Do all the points fall on this line? *See the Teacher's Answer Key. yes; yes*

d. Do negative values of h make sense in this situation? Explain. *See the Teacher's Answer Key.*

9. a. The cost, C, of a gift is the price, p, plus the sales tax on the price. If the sales tax is 5%, make a table of the cost of gifts that are priced at $10, $15, $20, $25, $30, $40, and $p. *See the Teacher's Answer Key.*

b. Graph C versus p on graph paper. *See the Teacher's Answer Key.*

c. Write an equation that relates C and p. Is it a linear equation? *$C = p + 0.05p$; yes*

10. a. The long distance cost for telephone calls at a certain time is given in the following table. Copy and complete the table.

Time (min):	1	2	3	4	5	6	7	x
Cost (cents):	25	43	61	79	97	*115*	*133*	*7 + 18x*

b. Write a rule for the cost of telephone service. *The cost equals 7 cents for the phone call plus 18 cents per minute.*

c. Graph the data on graph paper. Is the relation linear? *See the Teacher's Answer Key; yes*

d. What is the cost of a 15-minute call? What is the cost of a 17-minute call? *$2.77; $3.13*

11. Extension

a. How does the telephone company charge for long distance calls that are not an exact number of minutes? *Answers may vary. (The length of a call is usually rounded to the next higher whole number.)*

b. Does your graph in Exercise 10 fit this method of charging? *no*

Share & Summarize

c. Change your graph so that it gives the correct cost of a call whether it is a whole number of minutes or not. Be prepared to share your changes with the class. *See the Teacher's Answer Key.*

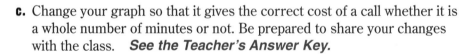
HOMEWORK PROJECT 1

Graphing Calculator Activity

You can learn how to use a graphing calculator to create tables in Activity 4 on page 61.

12. Extension

a. The expression $y = 10 - x$ defines a linear equation. Make a table of nine pairs of x and y values that satisfy the rule. *See the Teacher's Answer Key for a sample table.*

b. Graph the pairs. Do they lie on a line? If so, draw it. *Yes. See students' work.*

13. Extension

a. The expression $y = x - 10$ defines a linear equation. Make a table of nine pairs of x and y values that satisfy the expression. *See the Teacher's Answer Key for a sample table.*

b. Graph the pairs. Do they lie on a line? If so, draw it. *Yes. See students' work.*

14. Extension

    a. Are the expressions in Exercises 12 and 13 the same? Explain their similarities and differences.   ***No. Answers may vary. The same values for x produce y values with opposite signs.***

    b. Are the graphs the same? Explain their similarities and differences. ***See the Teacher's Answer Key.***

● PARTNER PROJECT 2

15. Extension

    a. For the data on male or female student athletes in Lindell High School on pages 5 and 6, are there any pairs of variables for one grade or for different grades whose points of the scatter plot should all lie on a line? Identify them. (Hint: Think about this. Do not try all sorts of combinations.) ***See the Teacher's Answer Key.***

    b. Use graphing technology to draw the median-fit lines for the pairs of data you identified in part a. Do the lines confirm your conjecture? Explain.   ***Answers may vary. For the age data, the conjectures will be confirmed.***

16. Extension

    a. Make a table of seven pairs of values of x and y that satisfy the expression $y = \frac{24}{x}$.   ***See the Teacher's Answer Key for a sample table.***

    b. Do the points lie on a line? Explain your answer.   ***No. Answers may vary.***

    c. Would a line make a good summarizing shape for these data?   ***no***

Portfolio Assessment

Select some of your work from this investigation that shows how you used a calculator or computer. Place it in your portfolio.

17. Extension

    a. For the expression given in Exercise 16, is there a value of y that corresponds to x when it is 0? Explain.   ***No. Dividing by 0 is undefined.***

    b. Can y be equal to 0? Explain.   ***No. There is no number that when divided into 24 will produce 0.***

● HOMEWORK PROJECT 2

18. Extension

    a. Describe the graphs of the linear equations $y = 5 + x$ and $y = x + 5$. ***Both equations produce the same graph.***

    b. Describe the graphs of linear equations $y = x + c$ and $y = c + x$ where c is an arbitrary constant number. ***Both equations produce the same graph.***

19. Extension

    Describe the graphs of the linear equations $y = x + 5$ and $y = x + 2$. ***The lines are parallel.***

Journal

20. Journal Entry Write a linear equation of your choice relating x and y. Make a coordinate system and graph ten pairs of points that satisfy your equation. Find the median-fit line for these points. What can you say about the median-fit line and the points in the scatter plot?   ***Answers may vary. See students' work. All points in the scatter plot will lie on the median-fit line.***

Linear Functions

When the rule relating one variable to a second is a linear equation, the relation is a **linear function**. Engineers use linear functions to model physical situations. Business people model economic situations using linear functions. The distance you travel on your bicycle going a constant speed of 10 miles per hour can be represented by a linear function.

Activity 5-1 Equations, Tables, and Graphs

Materials

graph paper

calculator

software

● GROUP PROJECT

1. a. Sofia Lopez sells new cars. Her monthly salary is $1,500. She also is paid $50 for every car she sells. How many cars do you think she could sell in a good month? In a poor month? *Answers may vary. (A good month might be 10 sales, a poor month 1 sale.)*

   b. What is Sofia's monthly income for her "good month" sales? her "poor month" sales? *Answers may vary. (A good month might be $2,000 in sales and a poor month $1,550.)*

2. a. Sofia's monthly income, y, is a linear function of the number of cars, x, she sells. It is represented by the linear equation $y = 50x + 1,500$. What does the 1,500 represent? What does the 50 represent? *See the Teacher's Answer Key.*

   b. Copy and complete the table below showing Sofia's monthly income as a function of the number of cars sold. How is the equation used to complete the table? *In place of x, put the appropriate x value in question and solve the equation for y.*

Number of Cars Sold: x	0	2	5	7	10	12	15	17	20
Sofia's Pay: y	1,500	1,600	1,750	1,850	2,000	2,100	2,250	2,350	2,500

   c. Make a scatter plot and find the median-fit line. *See the Teacher's Answer Key.*

d. Describe the relationship between the points and the line. ***All of the points lie on the line.***

e. How many cars does Sofia need to sell to earn $4000? ***50 cars***

Share & Summarize

f. If Sofia sold n cars in a month, how much does she earn? Be prepared to explain your reasoning to the class. ***1,500 + 50n dollars***

3. For each linear function, make a table of four pairs of values and then graph the ordered pairs on graph paper. Do all the points lie on a line? Create one more pair satisfying the equation. Graph it and tell whether it lies on the same line. ***See the Teacher's Answer Key for sample graphs. All five points in each case should lie on the line.***

a. $y = 5x + 2$ **b.** $y = -3x + 4$ **c.** $y = 1.5x - 2$

● PARTNER PROJECT

Use a graphing calculator or computer function grapher to investigate Exercises 4-8.

Graphing Calculator Activity

You can learn how to use a graphing calculator to plot points in Activity 5 on page 62.

4. a. Consider the equation $y = 2x + 1$. Copy and complete the table below by substituting the five different values of x into the right portion of the equation, and calculating the corresponding values of y. For example, if you let $x = -5$, then $2(-5) + 1$ is the value of y. In this case, $y = -9$. The first value in your table is shown.

x	−5	−2	1	3	5	7
y	−9	− 3	3	7	11	15

b. Use a graphing utility to plot the five points in the table above that satisfy the equation $y = 2x + 1$. ***See students' work.***

5. a. Examine the points plotted in Exercise 4. Do they appear to be on a line? ***yes***

b. Now have your graphing calculator or computer display the graph of the equation $y = 2x + 1$ without erasing the points earlier graphed. What is this graph? How are the points and the line related? ***The graph is a line through all of the previously-plotted points.***

Graphing Calculator Activity

You can learn how to use a graphing calculator to display a graph without erasing points graphed earlier and to trace a graph in Activity 6 on page 63.

6. a. Use the trace function of your graphing utility. Move the blinking cursor along the graph of $y = 2x + 1$ slowly. Do the coordinates of each of the points you graphed in Exercise 4 appear as coordinates of the trace point? ***No, the exact coordinates do not necessarily appear.***

b. Move the cursor along to another position and write down the coordinates of the point. Do the coordinates satisfy the equation? Illustrate with examples. ***Answers may vary. (Some choices will work out exactly, whereas some choices will introduce round-off error.)***

Journal

7. a. Journal Entry Describe how you would use the graph of the equation $y = 2x + 1$ to find a value of y corresponding to a value of x. *See the Teacher's Answer Key.*

b. Describe how you would use the equation to find a value of x corresponding to a known value of y. *The value of x is found by subtracting 1 from the y value and dividing by 2.*

8. For each equation, find the coordinates of three points that satisfy the equation, plot the three points using a graphing utility, and then draw the graph of the equation with the utility. Finally, indicate whether or not the points are on the graph.

a. $y = -2x + 1$ **b.** $y = 7x - 2$ **c.** $y = -4x + 3$

d. $y = 0.5x + 2$ **e.** $y = 1.5x - 3$ **f.** $y = x - 5$
See students' work; see the Teacher's Answer Key for graphs.

9. a. Choose the correct term to complete the following sentence.

The graph of an equation like $y = 3x - 2$ is (always, sometimes, never) a straight line. *always*

Share & Summarize

b. Be prepared to explain to the class why you made the choice you did in part a. *The equation y = 3x – 2 is linear, which implies that all points satisfying the equation will lie on the same line.*

● HOMEWORK PROJECT

10. Extension
a. What do you suppose the graph of the equation $x = 2y - 1$ looks like? *a straight line*
b. How would you construct the graph? Describe your method and then use it to draw the graph. What is the result? *Answers may vary.*

11. Extension
a. What do you suppose the graph of the equation $3x - 2y = 5$ looks like? *a straight line.*
b. How would you construct the graph? Describe your method and then use it to draw the graph. What is the result? *See the Teacher's Answer Key.*

Journal

12. Journal Entry Investigate the graphs of equations of the form $Ax + By + C = 0$, where A, B, and C are real numbers and not both A and B are 0. Make a conjecture concerning the nature of the graphs, and then write an argument you would use with a friend to convince him or her that your conjecture is correct. *Answers may vary. Ax + By + C = 0 is the standard form of a linear equation. Provided both A and B are not 0, the graph is a line.*

Materials

graph paper

colored pencils

software

 Graphing Calculator Activity

You can learn how to use a graphing calculator to display a family of linear functions in Activity 7 on page 64.

 **Share & Summarize**

You may use whatever means available to you, such as graph paper and pencil, computer software, or graphing calculator, to explore these questions.

● **PARTNER PROJECT**

1. The graphs of functions like $y = 2x$ and $y = 7x$ are lines.

   a. Explore the family of linear functions with equations $y = mx$ where m is positive. Graph $y = mx$ using six different values between $\frac{1}{10}$ and 6 for m. Look for patterns in the resulting graphs. Display your graphs on the same coordinate axes. ***See the Teacher's Answer Key.***

   b. Explore the family of linear functions with equations $y = mx$ where m is negative using the same method as in part a. ***See students' work.***

   c. Make a conjecture about the role of m in the graph of $y = mx$. Compare your observations with those of a classmate. Be prepared to share your reasoning with the class. ***Students should discover that m determines the steepness and direction of the line.***

2. Sketch a line and give an approximate value of m for a line containing the origin and lying: ***Answers may vary.***

   a. between $y = x$ and the x-axis
   Sample answer: m = $\frac{1}{2}$

   b. between $y = x$ and the y-axis
   Sample answer: m = 2

   c. between $y = -x$ and the x-axis
   Sample answer: m = $-\frac{1}{2}$

   d. between $y = -x$ and the y-axis.
   Sample answer: m = -2

3. a. Explore the family of linear functions with equations $y = x + b$ by varying the value of b through both positive and negative values and looking for patterns in the resulting graphs. Display your graphs on the same coordinate axes. ***See the Teacher's Answer Key for sample graphs.***

   b. Explore the family of linear functions with equations $y = 4x + b$. Use several different values for b. Display your graphs on the same coordinate axes. ***See the Teacher's Answer Key for sample graphs.***

   c. Explore the family of linear functions with equations $y = -2x + b$. Use several different values for b. Display your graphs on the same coordinate axes. ***See the Teacher's Answer Key for sample graphs.***

   d. Make a conjecture about the role of b in the graph of $y = mx + b$. Compare your conjecture with that of a classmate. ***Students should discover that b determines the point where the line crosses the y-axis.***

4. Use your conjectures about m and b to describe the graph of each of the following equations. If possible, use a graphing utility to verify your descriptions.   ***See the Teacher's Answer Key.***

a. $y = 5x - 6$ **b.** $y = -x + 0.4$

c. $y = 0.4x + 0.2$ **d.** $y = -10x - 8$

Share & Summarize

5. Journal Entry Summarize the results of the previous exercises. Be prepared to share your results with the class.
  a. For a line with equation $y = mx + b$, what does m determine?   ***See the Teacher's Answer Key.***
  b. For a line with equation $y = mx + b$, what does b determine?   ***b determines where the graph intersects the y-axis.***

6. For a line with equation $y = mx + b$, m is the **slope**, and b is the **y-intercept.**
  a. What is the slope and y-intercept of the graph of $y = 5x + 8$?   ***slope = 5; y-intercept = 8***
  b. Write the equation of a different line that has slope 5.   ***Sample answer: y = 5x – 3.***
  c. Graph the equations in parts a and b on the same coordinate axes. How do the lines appear to be related?   ***See the Teacher's Answer Key for graphs. The lines are parallel.***
  d. Write the equation of a line different from that in part a that has a y–intercept of 8.   ***Sample answer: y = 2x + 8.***

  e. Graph the equations in parts a and d on the same coordinate axes. How are the lines related?   ***See the Teacher's Answer Key for graphs. The lines intersect at (0,8).***

7. a. Write the equation of a line that has slope of 1 and intersects the two axes at the same point.   ***y = x***

  b. Draw the graph of your equation in part a.   ***See the Teacher's Answer Key.***
  c. Write an equation of a line that is steeper than that of the line in part a and that has a y-intercept of 0. Draw the graph of this equation on the same coordinate axes used for part b.   ***Sample answer: y = 2x. Look for m > 1 and b = 0. See the Teacher's Answer Key for graphs.***
  d. Write an equation of a line that is less steep than that of the line in part a and that has a y-intercept of 0. Draw the graph of this equation on the same coordinate axes used for part b.

  Sample answer: $y = \frac{1}{3}x$. Look for 0 < m < 1 and b = 0. See the Teacher's Answer Key for graphs.

● GROUP PROJECT

8. Extension
  a. What are the coordinates of the points where the graph of $3x - 2y = 6$ crosses the axes?   ***(0,– 3) and (2,0)***

b. A **term** is a number, a variable, or a product or quotient of numbers and variables. Some examples of terms are 5, $\frac{ab}{4}$, and $7k$. The **coefficient** is the numerical part of a term. For example, in $7k$, the coefficient is 7. In $\frac{ab}{4}$ the coefficient is $\frac{1}{4}$. How are the coordinates in part a related to the coefficients in the equation?   *See the Teacher's Answer Key.*

9. Extension

a. Describe the graph of the equation $\frac{x}{3} + \frac{y}{4} = 1$.   *a straight line*

b. At what points does the graph of $\frac{x}{3} + \frac{y}{4} = 1$ cross the x- and y-axes? How are the coordinates of these points related to the constants in the original equation?   *See the Teacher's Answer Key.*

10. Extension The graphs of the equations $y = 3x - 4$ and $y = -\frac{1}{3}x + 4$ are related in a special way. Investigate them and describe how they are related. How are the values of m in the equations related?   *The lines are perpendicular. The values of m are negative reciprocals of each other.*

Activity 5-3 Calculating Slopes and Finding Equations

Materials

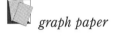
graph paper

Ski hills are often called "the slopes" because they have a large slope. A ski hill is shown in the figure below. The bottom of the hill has been given coordinates (0, 0). Two other points on the line of the hill have been marked also. These are the points A(400, 300) and B(600, 450). Is this a steep hill, or is it gentle?   *See Exercise 4d.*

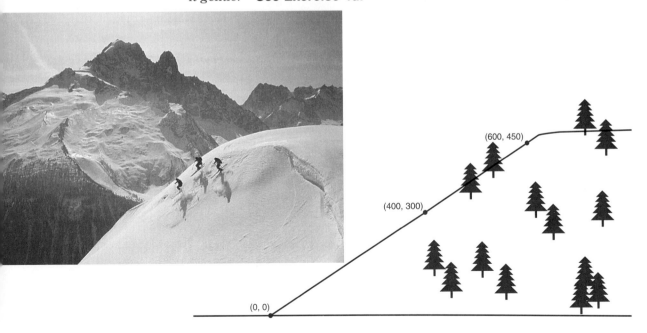

(600, 450)

(400, 300)

(0, 0)

1. a. The table below contains ten points on the line with equation $y = 2x + 3$. Copy and complete the table.

x	−4	−3	−2	−1	0	1	2	3	4	5
y	−5	−3	−1	1	3	5	7	9	11	13

b. Compare the differences between pairs of *y*-coordinates with the differences between corresponding pairs of *x*-coordinates. What pattern do you see?   **The y-coordinate differences are twice the x-coordinate differences.**

c. What is the slope of the line?   **2**

d. Is the slope related to the pattern you found in part b? How?   **See the Teacher's Answer key.**

2. a. The table below contains eight points on the line with equation $y = 3x - 4$. Copy and complete this table.

x	−1	0	1	2	3	4	5	6
y	−7	−4	−1	2	5	8	11	14

b. Refer to the table in part a. Compare the differences between pairs of *y*-coordinates with the differences between pairs of corresponding *x*-coordinates. What pattern do you see?   **See the Teacher's Answer key.**

c. What is the slope of the line?   **3**

d. Is the slope related to the pattern you found in part b? How?   **See the Teacher's Answer key.**

3. a. The table below contains eight points on the line with equation $y = \frac{3}{2}x + 1$. Copy and complete this table.

x	−4	−2	0	2	4	6	8	10
y	−5	−2	1	4	7	10	13	16

b. Compare the differences between pairs of *y*-coordinates with the differences between pairs of corresponding *x*-coordinates. What pattern do you see?   **As the y-coordinate changes by 3, the x-coordinate changes by 2.**

c. What is the slope of the line? $\frac{3}{2}$

d. Is the slope related to the pattern you found in part b? How?   **Yes. The slope is the same as the difference in y's divided by the differences in x's.**

4. a. Combine the information you found in Exercises 1-3 to make a conjecture about a way to compute the slope of a line from two points on the line.   *See the Teacher's Answer Key.*

b. Compare your conjecture with a classmate. Work out any differences to develop a joint conjecture.   *Answers may vary.*

c. Use your conjecture to find the slope of the ski slope on page 45. $\frac{3}{4}$

Share & Summarize

d. Is the slope steep or gentle? Be prepared to explain your reasoning to the class.   *Answers may vary. Students may decide that the slope is gentle because it is less than 1.*

5. Find the slope of the line containing each pair of points below.

a. (3, 4) and (5, 7)   **b.** (−1, 4) and (5, 4)   **c.** (−4, −3) and (−1, 2)
$\frac{3}{2}$ *or 1.5*   *0* $\frac{5}{3}$ *or 1.67*

d. (2, 5) and (−2, −2)   **e.** (3, 3) and (−2, −3)   **f.** (2, 4) and (3, 2)
$\frac{7}{4}$ *or 1.75* $\frac{6}{5}$ *or 1.2*   *−2*

6. For the lines given below, identify the coordinates of three points on the line. Calculate the slope of the line twice using different pairs of the three points. What result do you expect for the calculations for a given line?

a. $y = 2x - 1$ **b.** $y = -3x - 1$ **c.** $y = -4x - 9$
All pairs of points on each line produce the same slope.

Journal

7. Journal Entry Would you agree or disagree with this statement: "If I choose two points on a line and calculate the slope and you do the same with two other points on the line, the values for the slopes we get will be identical." Explain your position.   *See the Teacher's Answer Key.*

8. a. If a line has a negative slope, does it fall to the right or rise to the right on a coordinate system?   *it falls to the right*

b. What kind of line has a slope of 0?   *A horizontal line has a slope of 0.*

● GROUP PROJECT

Since two points determine a line and a line has a unique slope given by m in the equation $y = mx + b$, you can write the equation of a line whenever you know two points.

9. a. Suppose you know that the points (1, 1) and (4, 7) lie on a line. Devise a way to write the equation of the line containing these points in the form $y = mx + b$. (Remember that m is the slope of the line and that b is the y-intercept of the line.)   *Answers will vary.*

b. Summarize the procedure you developed. Compare your procedure with that of another group.   *See students' work.*

c. Are the procedures the same? Are they similar? Do they both work?   *Answers may vary.*

**Share &
Summarize**

d. Which procedure is easiest to do? Which method will you use? Be prepared to share your procedure with the class. ***Answers may vary.***

10. Use the procedure you developed in Exercise 9 to find the equation of the line containing each pair of points.

    a. (4, 4) and (5, 7)   **b.** (4, 2) and (5, 4)   **c.** (−4, −3) and (−1, 3)
       y = 3x − 8   ***y = 2x − 6***   ***y = 2x + 5***

    d. (2, 6) and (−2, −2)   **e.** (3, 3) and (−2, −3) **f.** (2, 4) and (3, 2)
       y = 2x + 2   ***y = 1.2x − 0.6***   ***y = −2x + 8***

11. Extension The y-intercept of a line is 4, and the point (−2, 2) is on the line. Find the equation of the line. ***y = x + 4***

12. Extension The y-intercept of a line is −2, and the x-intercept is 4. Find the equation of the line. ***y = 0.5x − 2***

13. Extension The slope of a line is −2, and the point (3, 5) is on the line. Find the equation of the line. ***y = −2x + 11***

14. Extension You read in the paper that a new highway will have a grade of 3%. What will be the slope of that road? ***See the Teacher's Answer Key.***

15. Extension Make a general argument, that, if a line has equation $y = mx + b$, then the slope of that line is m. ***See the Teacher's Answer Key.***

● HOMEWORK PROJECT

16. On a sheet of graph paper, sketch a line for each situation below. In each case, explain why there is only one such line or why there are many lines with the given characteristic(s). ***See students' graphs and explanations.***

**Portfolio
Assessment**

Select an item from your work that shows your creativity and place it in your portfolio.

    a. The line has a slope greater than 1. ***There are infinitely many lines with slopes greater than 1.***

    b. The line has a y-intercept of 3. ***There are infinitely many lines with a y-intercept of 3.***

    c. The line has a slope between 0 and 1 and a y-intercept of 2. ***See the Teacher's Answer Key.***

    d. The line has a slope less than −1 and a y-intercept of 0. ***See the Teacher's Answer Key.***

    e. The line contains the points (1, −2) and (3, 4). ***Two distinct points determine a line.***

    f. The line has a slope between 0 and −1. ***There are infinitely many lines with slopes between 0 and −1.***

    g. The line has a slope of −3 and a y-intercept of 2. ***There are infinitely many lines with y-intercept of 2. Of these lines, there is exactly one with slope of −3.***

Using Linear Functions

Median-fit lines are useful in exploring paired data such as that of Athletic Director Molar's student data. A helpful final technique is to write an equation of the line used to summarize those data. With this equation, you can substitute an initial value and get the predicted value immediately.

Activity 6-1   Equations for Fitted Lines

Materials

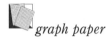
graph paper

calculator

software

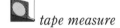
tape measure

🔵 GROUP PROJECT

1. Select 12 or 13 classmates and measure their height and stride lengths. Record your measurements in a table like the one shown below. You may need to convert inches to centimeters by multiplying by 2.54. ***See students' work.***

Name	Height (cm)	Stride Length (cm)

2. a. Make a scatter plot of stride length versus height.   ***See students' work.***

  b. Is the pattern of the data linear enough to be summarized by a line? Are there any outliers (made by classmates who did not walk naturally)? If so, reject those points, and construct the median-fit line for the graph. ***See students' work.***

3. a. Determine the equation of the median-fit line by determining the slope and the *y*-intercept.   ***See students' work.***

  b. Is the slope positive or negative? What is the slope?   ***Positive. Slopes will vary.***

  c. What does the slope represent in terms of the data?   ***The slope represents changes in stride length as height changes.***

  d. Is the *y*-intercept positive or negative? Estimate the *y*-intercept from the graph.   ***Positive. Answers will vary.***

  e. What does the *y*-intercept represent in terms of the data? Does it make sense?   ***The y-intercept represents the stride length of a person 0 cm tall; No.***

4. Use the graph of your line to predict the length of stride for a person with the given height.   *See students' work.*

   a. 150 cm   **b.** 160 cm   **c.** 170 cm   **d.** 180 cm

5. a. Use statistics software to produce a scatter plot and the median-fit line. *See students' work.*
   b. Compare the equation of the line produced by the software with your equation. How do they differ? Are the slopes similar? Are the *y*-intercepts close? *Answers may vary. Slopes and y-intercepts should be close.*

6. a. Use the data for the entire class and the statistics software to produce a scatter plot and summarizing line. What is the equation of this line? *See students' work.*
   b. Use this line to predict the stride length associated with each of the four heights in Exercise 4. Be prepared to share your findings with the class. *Answers will vary.*

Share & Summarize

7. a. Use the data in the three tables below to make scatter plots for Winning time versus Year. *See the Teacher's Answer Key.*

Women's 400-meter Freestyle Relay, 1912-1988

Year	Country	Time
1912	Great Britain	5:52.80
1920	United States	5:11.60
1924	United States	4:58.60
1928	United States	4:47.60
1932	United States	4:38.00
1936	Holland	4:36.00
1948	United States	4:29.20
1952	Hungary	4:24.40
1956	Australia	4:17.10
1960	United States	4:08.90
1964	United States	4:03.80
1968	United States	4:02.50
1972	United States	3:55.19
1976	United States	3:44.82
1980	East Germany	3:42.71
1984	United States	3:43.43
1988	East Germany	3:40.63

Source: Information Please Almanac, 1994

Women's 100-meter Butterfly, 1956-1988

Winner	Time
1956: Shelly Mann, United States	1:11.00
1960: Carolyn Schuler, United States	1:09.50
1964: Sharon Stouder, United States	1:04.70
1968: Lynn McClements, Australia	1:05.50
1972: Mayumi Aoki, Japan	1:03.34
1976: Kornelia Ender, East Germany	1:00.13
1980: Caren Metschuck, East Germany	1:00.42
1984: Mary T. Meagher, United States	59.26
1988: Kristin Otto, East Germany	59.00

Source: Information Please Almanac, 1994

Men's 400-meter Relay, 1912-1988 (Track)

Year	Country	Time (sec)
1912	Great Britain	42.40
1920	United States	42.20
1924	United States	41.00
1928	United States	41.00
1932	United States	40.00
1936	United States	39.80
1948	United States	40.60
1952	United States	40.10
1956	Germany	39.50
1960	United States	39.50
1964	United States	39.00
1968	United States	38.20
1972	United States	38.19
1976	United States	38.33
1980	Soviet Union	38.26
1984	United States	37.83
1988	Soviet Union	38.19

Source: Information Please Almanac, 1994

b. Find the equation of the summarizing line for each scatter plot. *See the Teacher's Answer Key.*

c. Use the equations of part b to predict the 1992 winning time for each event. *See the Teacher's Answer Key.*

d. Compare your predictions with the actual 1992 winning times. *Answers may vary. Actual winning times in 1992 are as follows: Women's 400-meter freestyle relay, 3:39.46; Womens' 100-meter Butterfly, 58.62; Men's 400-meter relay, 37.40.*

8. a. What is the slope of each summarizing line in Exercise 7? ***See the Teacher's Answer Key.***

b. Describe what the slope tells you about winning times as the years change. ***The fact that the slope is negative indicates that as years pass, winning times generally decrease.***

c. Where does the summarizing line cross the *x*-axis? What does this point represent in terms of winning time and year? Does this make sense? Explain. ***See the Teacher's Answer Key.***

● PARTNER PROJECT

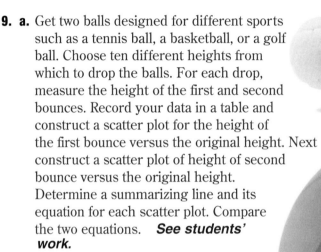

9. a. Get two balls designed for different sports such as a tennis ball, a basketball, or a golf ball. Choose ten different heights from which to drop the balls. For each drop, measure the height of the first and second bounces. Record your data in a table and construct a scatter plot for the height of the first bounce versus the original height. Next construct a scatter plot of height of second bounce versus the original height. Determine a summarizing line and its equation for each scatter plot. Compare the two equations. ***See students' work.***

b. Predict the bounce heights, given initial heights of 40 cm, 50 cm, 60 cm and 90 cm. ***Answers may vary.***

c. Can you predict the height of the second bounce from the scatter plot for first bounce versus original height? Explain. ***See the Teacher's Answer Key.***

d. Compare the lines generated for the two different balls. Are they similar? Explain. ***Both lines will have positive slopes.***

Graphing Calculator Activity

You can learn how to use a graphing calculator to find equations of summarizing lines in Activity 8 on page 65.

10. Extension

a. Find an equation of a summarizing line for twelfth grade 40-yard dash time versus ninth grade 40-yard dash time for males or females from the data on pages 5 and 6. ***males, y = 0.75x + 1.14; females, y = x + −0.28***

b. If each twelfth grade time was decreased by 0.1 second, what would be the equation of the summarizing line for these data? How is it related to the equation of the original line? ***See the Teacher's Answer Key.***

c. Increase each time in the twelfth grade by 0.2 second, and find the equation of the summarizing line. How is it related to the equations of the other lines? ***See the Teacher's Answer Key.***

d. Make a conjecture about the effects of adding or subtracting a constant to or from each *y*-coordinate value in relation to the equation of the summarizing line.   *See the Teacher's Answer Key.*

11. Extension

a. Using the same data you used in Exercise 10a, add 0.1 second to each ninth-grade time, and find the equation of the summarizing line for the resulting scatter plot.   *males, y = 0.75x + 1.24; females, y = x + −0.38*

b. Now subtract 0.2 second from each ninth-grade value and find the equation of the summarizing line for the resulting scatter plot.   *males, y = 0.75x + 0.94; females, y = x + −0.08*

c. Describe the effect on the summarizing line when the values of each *x*-coordinate are changed by adding a constant.   *See the Teacher's Answer Key.*

12. Extension

a. Given an equation $y = mx + b$, how is the graph of $y + c = mx + b$ related to the graph of $y = mx + b$? Verify your conjecture.   *See the Teacher's Answer Key.*

b. How is the graph of $y - c = mx + b$ related to the graph of $y = mx + b$? Verify your conjecture.   *See the Teacher's Answer Key.*

13. Extension

a. Given an equation $y = mx + b$, how is the graph of $y = m(x + c) + b$ related to the graph of $y = mx + b$? Verify your conjecture.   *See the answer to Exercise 11c.*

b. How is the graph of $y = m(x - c) + b$ related to the graph of $y = mx + b$? Justify your conjecture.   *The graph will shift right c units if c > 0, left c units if c < 0, and remain the same if c = 0.*

Activity 6-2   Manipulating Linear Equations

Materials

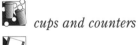
cups and counters

software

Often when working with linear equations you know a value for *y* and want to find a value for *x* . A procedure for dealing with this kind of situation is to model equations using cups and counters. In this model, a cup represents the variable, yellow counters ($\oplus$) represent positive integers, and red counters ($\ominus$) represent negative integers.

A model for the expression $3x + 2$ is shown at the right.

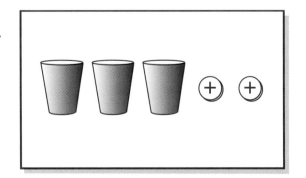

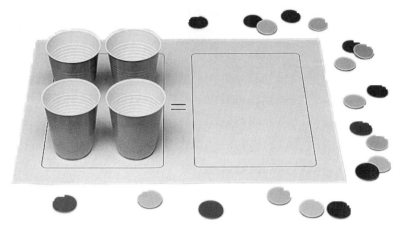

PARTNER PROJECT

1. Model each expression with cups and counters.

    a. $2x + 5$ **b.** $4x + 7$
    See the Teacher's Answer Key.

    c. To model $2x - 3$, first rewrite $2x - 3$ as $2x + (-3)$. Then $2x + (-3)$ can be modeled as follows.

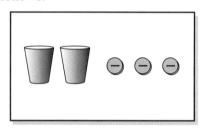

2. Model $3x - 5$ with cups and counters.   *See the Teacher's Answer Key.*

3. What expression is modeled by each model below?

    a.

    4x + 1

    b.

    x − 5

    c.

    1 + −1

    d.

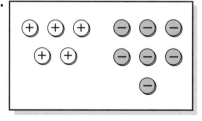

    5 + (−7)

4. Cups and counters can also be used to model equations. The equation $2x + 5 = 9$ is modeled below.

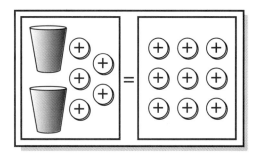

    a. If you replaced each cup with a positive counter, would the left side still equal the right side? Explain. ***No. You would then have 7 = 9 which is not true.***

    b. If you replaced each cup with a negative counter, would the left side still equal the right side? Explain. ***No. You would then have 3 = 9 which is not true.***

    c. Think of the cups as being weights on each side of a balance scale. How many counters would balance one cup? That is, determine the value of x for which this statement is true. Check your result by substituting for x in the original equation. ***Two counters balance one cup; x = 2; 2(2) + 5 = 9 which is true.***

    d. Describe the steps in your procedure. ***Answers may vary.***

    e. Use cups and counters to solve $3x + 4 = 13$. ***x = 3***

    f. Solve $5x + 3 = 18$ for x without using cups and counters. What was your first step? your second step? ***x = 3; Answers may vary.***

5. Cups and counters can also be used to model equations like $2x - 5 = 9$. First write $2x - 5 = 9$ as $2x + (-5) = 9$. Then model the equation as follows.

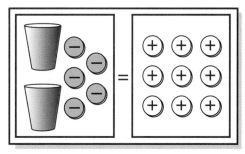

5a. Students may have difficulty with this problem; remind them that = is the same as a balance.

    a. Use the model to help you solve the equation. ***x = 7***

    b. Solve the equation $4x - 2 = 10$ using cups and counters. ***x = 3***

    c. Describe the steps you used in part b. Can you use your procedure to solve an equation without using cups and counters? Try it on $3x - 7 = 3$. ***Answers may vary.*** $x = \dfrac{10}{3}$

6. a. Compare the steps in your procedure for solving an equation like $3x - 5 = 23$ with those of others in your class. Are they the same? *See students' work.*

b. If the procedures are different, which way is easier for you to use? Which way is easier for you to understand? Do they both give the same result for x? If not, can they both be correct? *See the Teacher's Answer Key.*

Share & Summarize

c. Journal Entry Choose the procedure you understand most clearly for use in solving equations. Give a written description of its steps. Be prepared to share your reasons with the class. *See students' work.*

● HOMEWORK PROJECT

7. Copy each table below. Apply the procedure you chose in Exercise 6c to determine x for each value of y in the following equations.

a. $y = 3x - 3$ **b.** $y = -2x + 1$ **c.** $y = \frac{2}{3}x - 1$

x	y
2	3
$\frac{1}{3}$	-2
1	0
$1.0\overline{6}$	0.2
$0.8\overline{6}$	-0.4
1.25	$\frac{3}{4}$

x	y
-1	3
$\frac{3}{2}$	-2
$\frac{1}{2}$	0
0.4	0.2
0.7	-0.4
$\frac{1}{8}$	$\frac{3}{4}$

x	y
6	3
$-\frac{3}{2}$	-2
$\frac{3}{2}$	0
$\frac{9}{5}$	0.2
0.9	-0.4
$\frac{21}{8}$	$\frac{3}{4}$

8. a. Using the Lindell High School data on ninth- and twelfth-grade times in the 40-yard dash for males or females, found on pages 5 and 6, construct a scatter plot of twelfth-grade times versus ninth-grade times. *See the Teacher's Answer Key.*

b. Determine the equation of the summarizing line (median-fit). *males, y = 0.75x + 1.14; females, y = x + 0.28*

9. Use your equation from Exercise 8b to predict ninth-grade times given the following twelfth-grade times.

a. 5.1 s
males: 5.3 s; females: 5.4 s

b. 5.8 s
males: 6.2 s; females: 6.1 s

c. 4.9 s
males: 5.0 s; females: 5.2 s

d. 6.1 s
males: 6.6 s; females: 6.4 s

10. Use your equation from Exercise 8b to predict twelfth-grade times when given the following ninth-grade times.

a. 6.5 s
males: 6.0 s; females: 6.2 s

b. 6.0 s
males: 5.6 s; females: 5.7 s

c. 5.8 s
males: 5.5 s; females: 5.5 s

d. 5.3 s
males: 5.1 s; females: 5.0 s

GROUP PROJECT

11. a. Use the data on pages 5 and 6 to construct a scatter plot for twelfth grade bench-press (or leg-press) weight versus ninth grade bench-press (or leg-press) weight.   ***See the Teacher's Answer Key.***

Portfolio Assessment

Review items in your portfolio. Make a table of contents of the items, noting why each item was chosen. Replace any items that are no longer appropriate.

b. Determine the equation of the summarizing line.   ***leg-press, y = 1.29x + -13.33; bench-press, y = 0.67x + 87.22***

12. Use your equation from 11b to predict ninth-grade values for the following twelfth-grade values.

a. 180 lb
leg: 150 lb; bench: 139 lb

b. 210 lb
leg: 174 lb; bench: 184 lb

c. 165 lb
leg: 139 lb; bench: 117 lb

d. 250 lb
leg: 205 lb; bench: 244 lb

13. Use your equation from Exercise 11b to predict twelfth-grade values for the following ninth-grade values.

a. 120 lb
leg: 141 lb; bench: 167 lb

b. 145 lb
leg: 173 lb; bench: 184 lb

c. 160 lb
leg: 192 lb; bench: 194 lb

d. 100 lb
leg: 115 lb; bench: 154 lb

Journal

14. Journal Entry Describe how an equation of a summarizing line for a scatter plot, in the form $y = mx + b$, can be used:

a. to find y when x is given.   ***Substitute x into the equation, multiply it by m and add b.***

b. to find x when y is given.   ***Substitute y into the equation, subtract b from it, then divide by m.***

Graphing Calculator Activities

One way to determine whether there is a relationship between two sets of data is to display the information in a scatter plot. Graphing calculators are capable of drawing scatter plots for data that you enter into the memory.

Example **The elevation and number of clear days in 1992 for several U.S. cities are shown in the table below. Use a TI-82 graphing calculator to create a scatter plot of the data.**

City	Elev. (ft)	Clear days	City	Elev. (ft)	Clear days	City	Elev. (ft)	Clear days
Albany, NY	275	67	Dallas, TX	551	118	New Orleans, LA	4	83
Atlanta, GA	1010	110	Denver, CO	5283	113	Pittsburgh, PA	1137	55
Bismarck, ND	1647	90	Fresno, CA	328	170	Seattle, WA	400	67
Boston, MA	15	90	Nashville, TN	590	88	Springfield, MO	1268	103

Before you create a scatter plot, you must clear the statistical memories.

Enter: [STAT] 4 [2nd] [L1] [ENTER] [STAT] 4 [2nd] [L2] [ENTER]

Next, enter the data. Enter the elevations in list L1 and the clear days in list L2.

Enter: [STAT] [ENTER] *Accesses the statistical lists.*

275 [ENTER] 1010 [ENTER] ... 1268 [ENTER]

[▶] 67 [ENTER] 110 [ENTER] ... 103 [ENTER]

After the data is entered, the range for the graph must be set. A viewing window of [0, 5500] by [0, 200] with a scale factor of 500 on the x-axis and 25 on the y-axis is appropriate for this data.

Enter: [WINDOW] [ENTER] 0 [ENTER] 5500 [ENTER] 500 [ENTER] 0

[ENTER] 200 [ENTER] 25 [ENTER]

Now, you may choose the type of statistical graph and create the graph.

Enter: [2nd] [STAT PLOT]

● **Try This**

The table below shows the average heights in feet of 30 men and 30 women at different ages. Use a TI-82 graphing calculator to create a scatterplot for the data. *See students' work.*

Age (yr)	1	3	5	10	12	15	18	20	22
Men	2.4	3.2	3.8	4.5	4.8	5.3	5.7	5.9	6.0
Women	2.5	3.3	3.7	4.4	4.9	5.2	5.3	5.4	5.5

Graphing Calculator Activity 2: The Correlation Coefficient

When you are analyzing a set of data, it is often difficult to determine a relationship with a quick comparison. Finding the correlation coefficient of the data can be very helpful. You can use a graphing calculator to find the correlation coefficient of a set of data you enter.

Example The table below shows the rank in area and the rank in order of when each of the contiguous United States entered the Union. Find the correlation coefficient of the data.

State	Area	Order	State	Area	Order	State	Area	Order	State	Area	Order
AL	28	22	IA	24	29	NE	14	37	RI	48	13
AZ	5	48	KS	13	34	NV	6	36	SC	39	8
AR	26	25	KY	36	15	NH	43	9	SD	15	40
CA	2	31	LA	30	18	NJ	45	3	TN	33	16
CO	7	38	ME	38	23	NM	4	47	TX	1	28
CT	46	5	MD	41	7	NY	29	11	UT	10	45
DE	47	1	MA	44	6	NC	27	12	VT	42	14
FL	21	27	MI	22	26	ND	16	39	VA	35	10
GA	20	4	MN	11	32	OH	34	17	WA	19	42
ID	12	43	MS	31	20	OK	17	46	WV	40	35
IL	23	21	MO	18	24	OR	9	33	WI	25	30
IN	37	19	MT	3	41	PA	32	2	WY	8	44

Clear lists L1 and L2 in the statistical memory before you enter the data.

Enter: [STAT] 4 [2nd] [L1] [ENTER] [STAT] 4 [2nd] [L2] [ENTER]

Now enter the data. Enter the area in list L1 and the order in list L2.

Enter: [STAT] [ENTER] *Accesses the statistical lists.*

      28 [ENTER] 5 [ENTER] 26 [ENTER] ... 8 [ENTER]

      [▶] 22 [ENTER] 48 [ENTER] 25 [ENTER] ... 44 [ENTER]

The TI-82 graphing calculator is capable of finding many different correlation values. The correlation coefficient we will use is the Pearson-product moment correlation.

Enter: [STAT] [▶] 9 [ENTER]

The correlation coefficient for the area and order of the contiguous United States is -0.8030829353. The data has a very strong negative relationship.

● Try This

Find the correlation coefficient to the nearest hundredth. *0.95*

				Millions of Elementary and Secondary School Students					
Year	1900	1910	1920	1930	1940	1950	1960	1970	1980	1990
Students	10.6	12.6	16.2	21.3	22.0	22.3	32.5	42.5	38.2	38.0

Graphing Calculator Activity 3: Median-Fit-Lines

When data is collected in real-life situations, the relationship between two sets of values is rarely a straight line. However, the relationship may be approximated by a straight line. One way to approximate the graph of data is by finding the median-fit line. Your graphing calculator can find the median-fit line for data you enter.

Example Draw a scatter plot and the median-fit line for the data about the orbits of ten asteroids that is given in the table below.

Asteroid	Ceres	Pallas	Juno	Vesta	Astraea	Hebe	Iris	Flora	Metis	Hygeia
Mean Distance from Sun (millions of miles)	257.0	257.4	247.8	219.3	239.3	225.2	221.4	204.4	221.7	222.6
Orbital period (years)	4.60	4.61	4.36	3.63	4.14	3.78	3.68	3.27	3.69	5.59

Begin by clearing the statistical lists and entering the data.

Enter: STAT 4 2nd L1 ENTER STAT 4 2nd L2 ENTER
STAT ENTER 257.0 ENTER 257.4 ENTER … 222.6
ENTER ▶ 4.60 ENTER 4.61 ENTER … 5.59 ENTER

Enter the viewing window for the scatter plot. The data suggests a window of [200, 260] by [3, 6] with a scale factor of 10 for the *x*-axis and 0.5 for the *y*-axis.

Enter: WINDOW ENTER 200 ENTER 260 ENTER 10 ENTER 3
ENTER 6 ENTER 0.5 ENTER

Create the scatter plot by pressing 2nd STAT PLOT ENTER and then using the arrow and ENTER keys to highlight "On", the scatter plot, L1 as the Xlist, L2 as the Ylist, and • as the mark.

Use the CLEAR key to clear any equations that are already in the Y= list before you enter the equation of the median-fit line.

Find the equation of the median-fit line by pressing STAT ▶ 4 ENTER . Press Y= VARS 5 ▶ ▶ 7 to add the equation to the Y= list. Press GRAPH to see the scatter plot and the median-fit line.

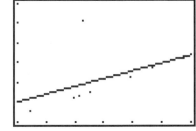

⬤ Try This

The table below shows the percent of American television owners who had cable television in different years. Create a scatterplot and graph the median-fit line for the data. *See students' work.*

Year	Percent	Year	Percent	Year	Percent	Year	Percent
1977	16.1	1981	28.3	1985	46.2	1989	57.1
1978	17.9	1982	35.0	1986	48.1	1990	59.0
1979	19.4	1983	40.5	1987	50.5	1991	60.6
1980	22.6	1984	43.7	1988	53.8	1992	61.5

Graphing Calculator Activity 4:   Tables

A graphing calculator is a powerful tool for studying functions. One way you can examine a function is to create a table of values. A TI-82 graphing calculator will allow you to create a large table of values quickly.

Example   As a thunderstorm approaches, you see lightening as it occurs, but you hear the accompanying sound of thunder a short time afterward. The distance y in miles, that sound travels in x seconds is given by the equation y = 0.21x. Use a graphing calculator to create a table of values for x = {0, 0.5, 1, 1.5, 2, 2.5, 3, 3.5, ...}. How far away is lightening when the thunder is heard 3 seconds after the light is seen?

First, enter the function $y = 0.21x$ into the Y= list. Press $\boxed{Y=}$ to access the list. Then enter the equation in as function Y1. Use the $\boxed{CLEAR}$ key to remove any equations that are already in the list.

Now press $\boxed{2nd}$ $\boxed{TblSet}$ to display the table setup menu. The table is to start at 0, so enter 0 as the TblMin value and press $\boxed{ENTER}$. ΔTbl is the change between each pair of successive x-values in the table. Enter 0.5 as the ΔTbl value and press $\boxed{ENTER}$. Use the arrow and keys to highlight "Auto" for both the dependent and independent variables so that the calculator will construct the table automatically. Press $\boxed{2nd}$ $\boxed{Table}$ to display the completed table.

Use the arrow keys to scroll through the table entries. According to the table, when thunder is heard 3 seconds after the lightening is seen, the lightening is 0.63 miles away.

X	Y₁	
0	0	
.5	.105	
1	.21	
1.5	.315	
2.	.42	
2.5	.525	
3	**.63**	

X = 3

● Try This

1. Geothermal energy is generated whenever water comes in contact with heated underground rocks. The heat turns the water into steam that can be used to make electricity. The underground temperature of rocks varies with their depth below the surface. The temperature, y, in degrees Celsius is estimated by $y = 35x + 20$, where x is the depth in kilometers.

    a. Use a graphing calculator to create a table of values for x = {0, 5, 10, 15, 20, 25, 30, 35, ...}. *See the Teacher's Answer key.*

    b. What would be the temperature of rocks that are 15 kilometers deep?
    545°

2. The distance, y, in feet that an object falls in x seconds is found by $y = 16x^2$.

    a. Use a graphing calculator to create a table of values for x = {0, 0.5, 1, 1.5, 2, 2.5, 3, 3.5, ...}. *See the Teacher's Answer key.*

    b. How far will an object fall in 7 seconds? *784 feet*

Graphing Calculator Activity 5:   Plotting Points

The graphics screen of a graphing calculator can represent a coordinate plane. The *x*- and *y*-axes are shown, and each point on the screen is named by an ordered pair. You can plot points on a graphing calculator just as you do on a coordinate grid.

The program below will plot points on the graphics screen. In order to use the program, you must first enter the program into the calculator's memory. To access the program memory, use the following keystrokes.

Enter: PRGM ▶ ▶ ENTER

Example   Plot the points (0, −4), (3, 1), (−5, −4), (−1, 6), and (8, 2) on a graphing calculator.

First, set the range. The notation [−10, 10] by [−8, 8] means a viewing window in which the values along the *x*-axis go from −10 to 10 and the values along the *y*-axis go from −8 to 8.

```
Prgm1: PLOTPTS
:FnOff
:PlotsOff
:ClrDraw
:Lbl 1
:Disp "X="
:Input X
:Disp "Y="
:Input Y
:Pt-On(X, Y)
:Pause
:Disp "PRESS Q TO QUIT,"
:Disp "1 TO PLOT MORE"
:Input A
:If A = 1
:Goto 1
```

Enter: WINDOW   ENTER

(−) 10 ENTER 10 ENTER 1 ENTER

(−) 8 ENTER 8 ENTER 1 ENTER

The program is written for use on a TI-82 graphing calculator. If you have a different type of programmable calculator, consult your User's Guide to adapt the program for use on your calculator.

Now run the program.

Enter: PRGM 1 ENTER

Enter the coordinates of each point. They will be graphed as you go. Press ENTER after each point is displayed to continue in the program.

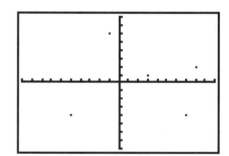

● Try This

Use the program to graph each set of points on a graphing calculator. Then sketch the graph. ***See students' work.***
  1. (7, −1), (−3, 6), (−1, 2), (6, 8)
  2. (−1.7, 2.2), (0.8, 1.9), (−1.2, 0.1), (−2.1, −3.7), (1.6, 3.2)
  3. (−32, 4), (25, −15), (−13, −18), (−5, −11)
  4. (92, 40), (−67, 21), (−51, −37), (24, −16), (32, −57), (89, 21)

Graphing Calculator Activity 6: Tracing a Function

You can explore the characteristics of different functions by observing their graphs. The first step to graphing a function on a graphing calculator is to set an appropriate range. A viewing window of $[-10, 10]$ by $[-10, 10]$ with a scale factor of 1 on both axes denotes the domain values $-10 \leq x \leq -10$ and the range values $-10 \leq y \leq 10$. The tick marks on both axes in this viewing window will be one unit apart. This is called the standard viewing window. The standard viewing window is a good place to start when graphing an unfamiliar function.

Example Graph $y = 2x + 3.7$ in the standard viewing window. Then use the trace function to determine whether the point $(-1, 2.3)$ is on the graph.

First, enter the function into the Y= list. If any functions are already on the list, clear them by using the arrow keys to move the cursor anywhere in the equation and then pressing the CLEAR key.

Enter: Y= 2 X,T,θ + 3.7

Now, select the standard viewing window and graph. The TI-82 graphing calculator has the standard viewing window as a choice on the zoom menu, so you can choose it without entering the range values manually.

Enter: ZOOM 6 *Selects the standard viewing window and completes the graph.*

You can use the trace function of a graphing calculator to find approximations of the coordinates of points that appear on the graph of a function. Press TRACE to get a blinking cursor on the graph. The right and left arrow keys allow you to move the cursor along the graph. Move the cursor to a point on the line where the *x*-coordinate is as close as possible to -1. Since the *y*-coordinate is not close to 2.3, we can determine that $(-1, 2.3)$ is not on the graph of $y = 2x + 3.7$.

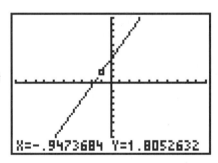

Try This

Graph each function in the standard viewing window. Then determine whether the given point is on the graph of the function.

1. $y = 3x - 2.1$; $(6, 6.2)$ *no*
2. $y = -2x - 4$ $(-3, 2)$ *yes*
3. $y = 6 - x$; $(7, -1)$ *yes*
4. $y = 2.5 + 3x$; $(-3, -6.5)$ *yes*
5. $y = 11x - 6.9$; $(1, 7.4)$ *no*
6. $y = 13 - 5.5x$; $(2.5, 1.5)$ *no*
7. $y = x + 5.93$; $(4, 9.93)$ *yes*
8. $y = -0.5x + 4.22$; $(8.44, 0)$ *yes*

Graphing Calculator Activity 7:   Families of Linear Equations

A family of graphs is a group of graphs that displays one or more similar characteristics. Many linear functions are related because they have the same slope or the same y-intercept as other functions in the family. All linear functions can be written in the form $y = mx + b$, where m represents the slope of the line and b is the y-intercept. You can graph several functions on the same screen and observe if any family traits exist.

Example   Graph the following functions on the same screen in the standard viewing window. Then describe the family of graphs to which they belong.

$y = 0.5x + 1$ $y = 3x + 1$

$y = x + 1$ $y = 5x + 1$

The TI-82 graphing calculator allows you to graph up to nine functions at one time. Enter each equation into the Y= list. *Be sure to clear any equations that are in the list before you start.*

Enter: Y= 0.5 X,T,θ + 1 ENTER X,T,θ + 1 ENTER

3 X,T,θ + 1 ENTER 5 X,T,θ + 1

The standard viewing window can be selected automatically from the zoom menu. Press ZOOM 6. The graphs will appear automatically.

When all the functions are graphed on the same screen, you can observe that they are all lines and that they all pass through the point (0, 1). However, their slopes are different. If you compare each graph with its slope, you find the greater the slope, the greater the angle formed by the line and the x-axis. This family of graphs is described as lines that have a y-intercept of 1.

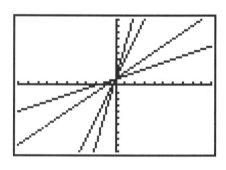

⬤ Try This

Graph the following functions on the same screen. Then describe the family of graphs to which they belong.   *See the Teacher's Answer Key.*

1. $y = 3x$
$y = 3x + 1$
$y = 3x - 2$
$y = 3x + 5$

2. $y = 4$
$y = -1$
$y = 2$
$y = 5.2$

3. $y = 2x - 4$
$y = 4x - 4$
$y = -3x - 4$
$y = -x - 4$

Graphing Calculator Activity 8: Finding the Equation of a Median-Fit Line

Graphing calculators are capable of finding the median-fit line for a set of data that you enter into the statistical memory. Once the equation of the median-fit line is found, you can use the equation to approximate solutions to other problems.

Example The table below shows median age of the resident population of the United States for different years. Find the equation of the median-fit line of the data. Then use the equation to predict the median age in the year 2010.

Year	Median Age	Year	Median Age	Year	Median Age
1820	16.7	1900	22.9	1970	28.0
1840	17.8	1920	25.3	1980	30.0
1860	19.4	1930	26.4	1983	30.8
1870	20.2	1940	29.0	1984	31.1
1880	20.9	1950	30.2	1985	31.4
1890	22.0	1960	29.5	1990	32.9

First, clear the statistical lists and enter the values. Use list L1 for the years and list L2 for the populations.

Enter: STAT 4 2nd L1 ENTER   STAT 4 2nd L2 ENTER
STAT ENTER 1820 ENTER 1840 ENTER … 1900 ENTER
▶ 16.7 ENTER 17.8 ENTER … 32.9

The TI-82 graphing calculator will find many different statistical equations. The median-fit line is one of the choices available on the statistical calculation menu.

Enter: STAT ▶ 4 ENTER

The calculator will display the values of a and b for a median-fit line of the form $y = ax + b$. To the nearest thousandth, the values for this data are $a = 0.094$ and $b = -155.245$. Thus, the equation of the median-fit line is $y = 0.094x - 155.245$. Substituting 2010 for x in the equation, gives an estimate of 33.695 for the median age in the year 2010.

Try This $y = 20.676x + 83.75$; 65.5%

Find the equation of the median-fit line. Then predict the percentage of calories from carbohydrates for a food with 27% calories from fat.

Food	% Calories from Fat	% Calories from Carbohydrates
Apple (medium)	9	89
Bagel (plain)	6	76
Banana (medium)	2	93
Bran Muffin (large)	40	53
Fig Newtons (4 bars)	18	75
PowerBar (any flavor)	8	76
Snickers bar (regular size)	42	51
Ultra Slim-Fast bar (one)	30	63

Glossary

A

Associated (p. 8) Two sets of data are associated if an increase in one is accompanied by an increase in the other or an increase in one is accompanied by a decrease in the other.

Example Two variables such as height and shoe size are associated because an increase in one variable accompanies an increase in the other.

C

Coefficient (p. 45) The numerical part of a term.

Example In 8*ab* the coefficient is 8.

Correlation (p. 13) An interdependence between sets of data.

Example The data represented in a scatter plot whose points either rise or fall from left to right have a correlation.

Correlation (p. 14) A number that indicates the degree of correlation between two coefficient sets of data.

Example A correlation coefficient of 0.9 represents a strong positive association, while a - 0.01 represents a weak negative association.

F

Fitting a line to the data (p. 16) Drawing a line to summarize a scatter plot so that there are about as many points above the line as below it and that the line contains some of the points.

Example

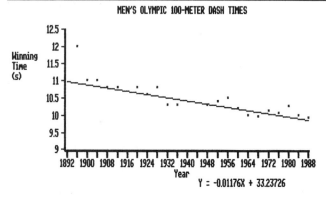

Data Insights: Scatter Plot

$Y = -0.01176X + 33.23726$

L

Linear function (p. 40) A relation whose rule relating one variable to a second is a linear equation.

Example The equation $y = 2x + 1$ represents a linear function.

M

Mathematical model (p. 17) A representation of data that can be used to predict one value of a pair of data when given the other value.

Example A visually-fit line is an example of a mathematical model for a pair of data.

Mean residual (p. 22) A measure of how well a summarizing line fits the data. It is the sum of the residuals of a summarizing line divided by the number of residuals of the line.

Example If the sum of the 12 residuals of a line is 6, the mean residual is 6 ÷ 12 or 0.5. A relatively small mean residual such as this indicates that a summarizing line is a good fit to the data.

Median-fit line (p. 24) A summarizing line that is drawn using three medians in a set of data.

Example

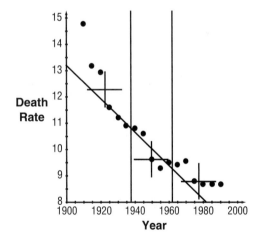

Median point (p. 26) A point whose x-coordinate is the median of the x-coordinates of all the points in a set of data and whose y-coordinate is the median of the y-coordinates of all the points in a set of data.

Example A set of data that contains the points (2, 3), (1, 0), (3, 4) has a median point of (2, 3).

N

Negative association
(p. 9)

The pattern on a scatter plot where the points get closer to the horizontal axis as the values on the horizontal axis increase.

Example

A scatter plot of *the value of a car versus the age of a car* will have a negative association.

No association (p. 9)

The pattern on a scatter plot where the points are scattered all over the graph and a reliable prediction cannot be made.

Example

A scatter plot of *height versus birth month* will have no association.

Perfect association
(p. 13)

A pattern on a scatter plot where all the points fall on a line.

Example

A scatter plot of *number of hours worked versus gross pay* will have perfect association.

Positive association
(p. 9)

The pattern on a scatter plot where the points get further away from the horizontal axis as the values on the horizontal axis increase.

Example

A scatter plot of *age versus height* will have positive association.

R

Residual (p. 22)

The vertical distance from a point to a summarizing line in a scatter plot.

Example

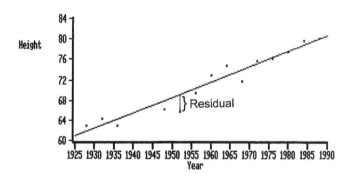

S

Scatter plot (p. 7)   A graph on a coordinate system that shows the relationship between two sets of data.

Example   The graph below is a scatter plot of *40-yd dash time versus body weight.*

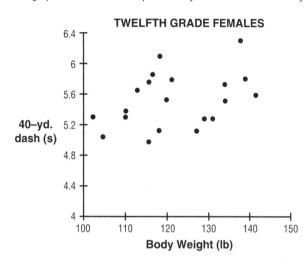

Slope (p. 44)   The slope of a line is the difference between the *y*-coordinates of any two points on the line divided by the difference between the corresponding *x*-coordinates. The slope of a line described by $y = mx + b$ is m.

Example   The slope of the equation of the line $y = 3x - 2$ is 3.

Spearman rank correlation (p. 15)   The correlation coefficient of ranked data that has no ties. It is found by using the following formula where *n* represents the number of rankings.

$$r = 1 - \frac{6(\textit{sum of the squares of the differences in the rankings})}{n(n^2 - 1)}$$

Example

My Ranking	Brian's Ranking	Difference	Difference^2
1	10	9	81
2	1	-1	1
4	8	4	16
6	3	-3	9
8	2	-6	36
7	4	-3	9
3	5	2	4
9	9	0	0
10	6	-4	16
5	7	2	4

$$r = 1 - \frac{6(81 + 1 + 16 + 9 + 36 + 9 + 4 + 16 + 4)}{10(100 - 1)} = \frac{1056}{990} = 1.067$$

Summarizing line
(p. 16)

A line that follows the general pattern of a scatter plot.

Example

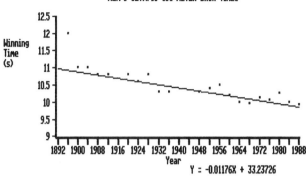

T

Term (p. 45)   A number, a variable, or a *product or quotient* of numbers and variables.

Example   Some examples of terms are 2, $\frac{x}{3}$, and 7*ab*.

V

Visually-fit line
(p. 17)

A visual estimate of a summarizing line for a set of data.

Example   A line drawn by placing a piece of string on a scatter plot and using a ruler to draw the line is a visually-fit line.

Y

y-intercept (p. 44)   The value of a function when *x* is 0. It is the point where the graph of an equation crosses the *y*-axis.

Example   The *y*-intercept of the equation $y = 8x + 5$ is 5.

Index

Photo Credits